T0365847

ISBN: Softcover 978-1-5245-7555-7
EBook 978-1-5245-7556-4

Print information available on the last page

Rev. date: 01/20/2017

To order additional copies of this book, contact:
Xlibris
1-888-795-4274
www.Xlibris.com
Orders@Xlibris.com

PRE - CALCULUS
For College Students

Alice Gorguis
North Park University
Chicago, Illinois

About the Author

Alice Gorguis, a faculty member of Mathematics department at North Park University in Chicago, received her PhD in Applied Mathematics from University of Illinois at Chicago Circle (UIC), and 68-credit hours toward PhD in Solid State Physics, and received M.S in Physics from Northeastern Illinois University in Chicago (NEIU), also received M.S in Statistics from Northeastern Illinois University in Chicago. Worked as a full time faculty for 18 years.

As author and reviewer of Elsevier Publishing Journal, she authored and published 10 articles, and reviewed over 40 articles for publication for many different Elsevier Journals

She published the following books with Xlibris publishing:

- Intermediate Algebra with Analytic Geometry. ISBN: 978-5245-2347-3 (7/30/2016)
- Statistics Tools / 2nd Edition. ISBN: 978-1-5144-3769-8 (01/20/2016).
- Real Analysis / 3rd Edition. ISBN: 978-1-5035-8974-2 (8/10/15).
- Statistics Tools /1st Edition. ISBN: 978-1-5035-7970-5 (7/13/15).
- Vector Calculus / 3rd Edition. ISBN: 978-1-5035-8039-8 (7/10/15).
- Ordinary Differential Equations /1st Edition. ISBN: 978-1-4990-6036-2 (11/12/2014).
- Vector Calculus for college students / 2nd Edition. ISBN: 978-1-4990-4891-9 (7/30/14).
- Vector Calculus for college students / 1st Edition. ISBN: 978-1-4836-7257-1 (8/6/13)
- Real Analysis a step by step approach / 2nd Edition. ISBN: 978-1-4797-8459-2 (2/20/13).

Contents

Preface

This book is designed for students who need additional preparation for math classes, at North Park University, numbered 1510 Calculus-I, and 1520 Calculus-II or higher. The book covers all the topics that are required for this course: topics in properties and graphs of functions such as: polynomial functions, rational functions, exponential functions, logarithmic functions, trigonometric functions, and inverse trigonometric functions; Solving equations, and inequalities; trigonometric identities, and applications of trigonometric functions; complex numbers; and conic sections

For some students this is the last course they need to take, for others just the first in series of many, depending on students' major.

The text explains the method of solving problems step by step, and shows more than one method to solve some problems, also shows how to solve problems graphically and using technology (TI) so the student will learn how to solve the problem algebraically, then see the solution graphically.

Most of the chapters are ended with application section to learn how to apply the topic to real life situation. Also, at the end of each chapter there is a chapter exercise, and a chapter test, students can use the chapter exercise as a study guide for their tests.

My goal of writing this book is to make it easy for students to read through each page and doesn't overwhelm, or complicate the covered topic, but gives the skills and practice that they need.

Best Wishes

Alice Gorguis

January '2017

1. Functions

Chapter – 1
Functions

Objectives: 1.1 Introduction to Functions
1.2 Properties of Functions
1.3 Average Rate of Change of Functions
1.4 Piecewise Defined Functions
1.5 Transformation of Functions

1.1 Introduction to Functions

Function

Function is a special type of relation, that correspondence between x-values or the Domain and y-values or the Range.

In the previous level of Algebra we studied equations which are defined in general as:
$y = m x + b$, for first order, or $ax^2 + bx + c = 0$, for second order…. etc, but now we are using a different terminology called **"Functions".**
To distinguish between equations and functions, graphically, we use the **vertical line test** (the Red Line Test): The test line cross the function graph in one single point only.

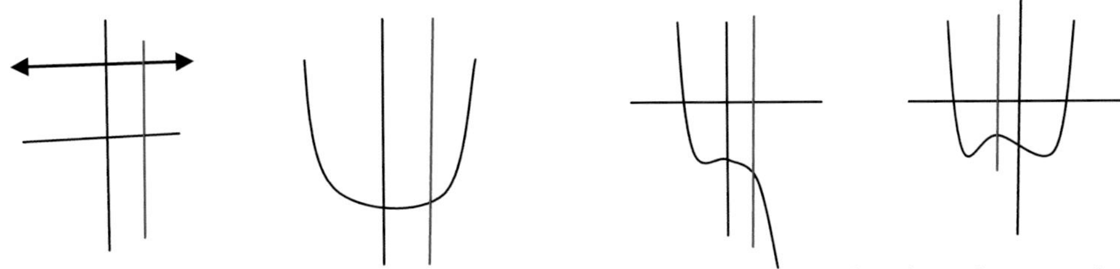

If the test line cross the graph of the equation in one single point only, then the equation is considered to be a function, and can carry a new label f(x) where y= f(x). This means the function can be written as y or f(x), but the equation that is not a function cannot be written as f(x).The following equations are not functions, because the test-line (the red line) is crossing the graph in more than one point:

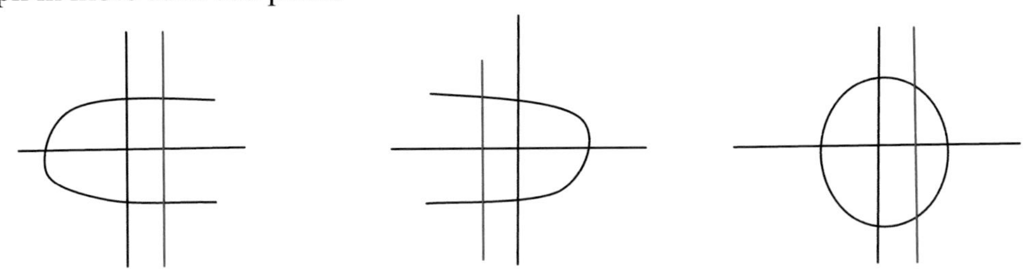

Example-1	Ordered Pairs as Relation

Determine whether each relation is a function? If function, state the domain and range.

A.

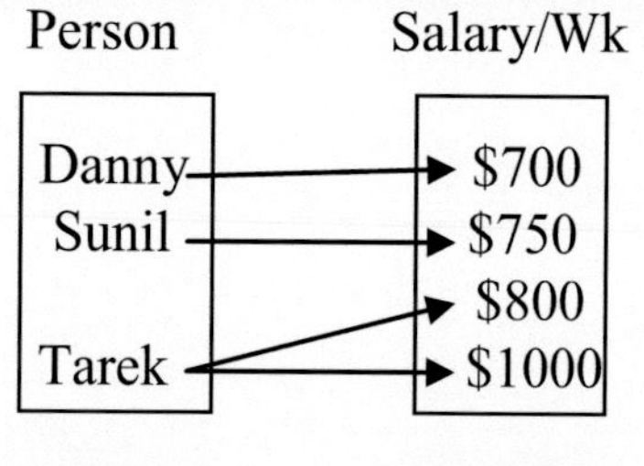

B.

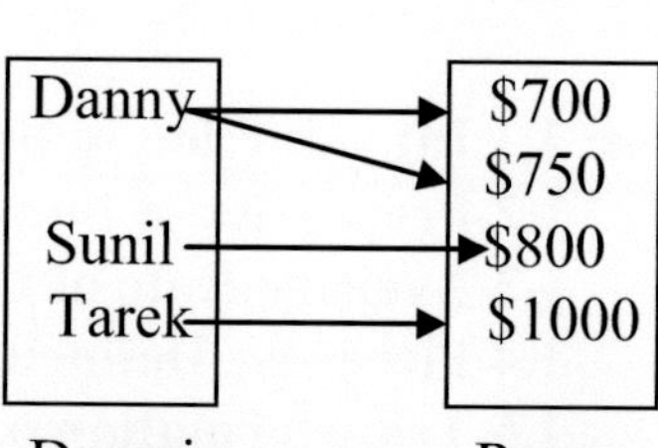

C.

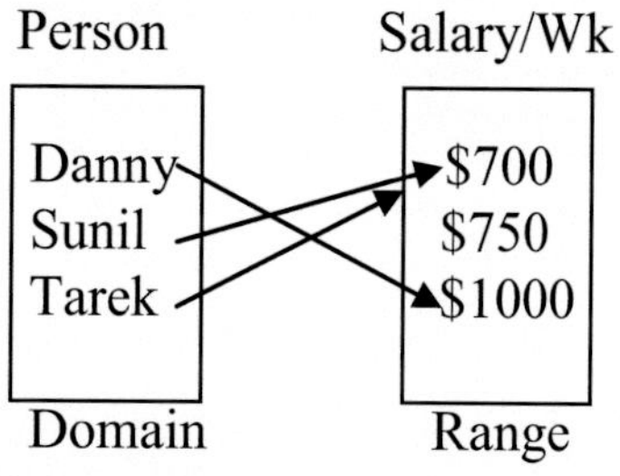

D.

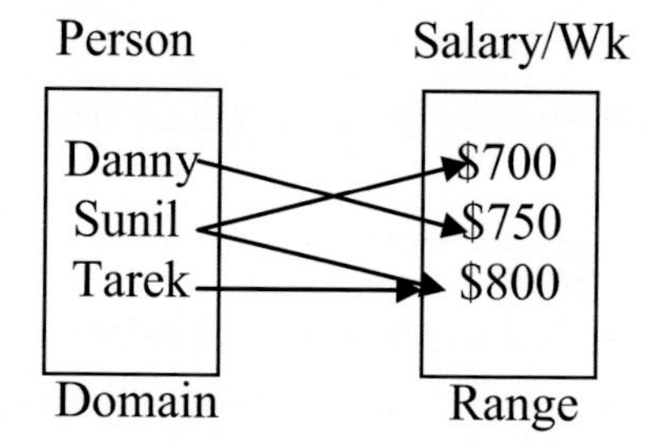

Solution:

A. Not a function.

B. Not a function

C. Function with Domain={Danny, Sunil, Tarek}, and Range={$700, $750, $100}

D. Not a function

Functions can also be defined as:

a. For every x-value there is only one y-value:
The line between the points is a function

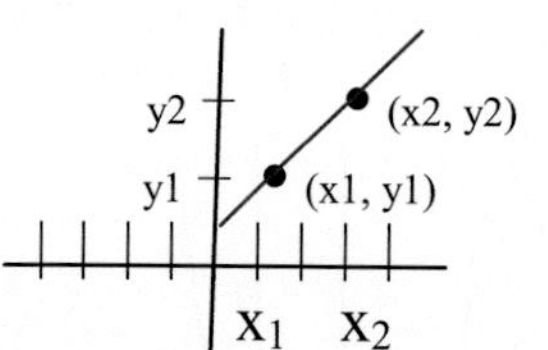

b. For every y-value there is more than one x-value:

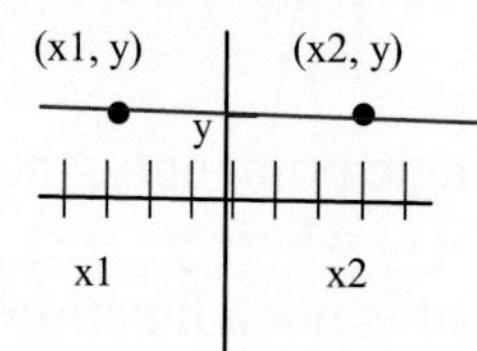

The line between the points is a function

Non-Functions are defined as: for every x-value there is more than one y-value:

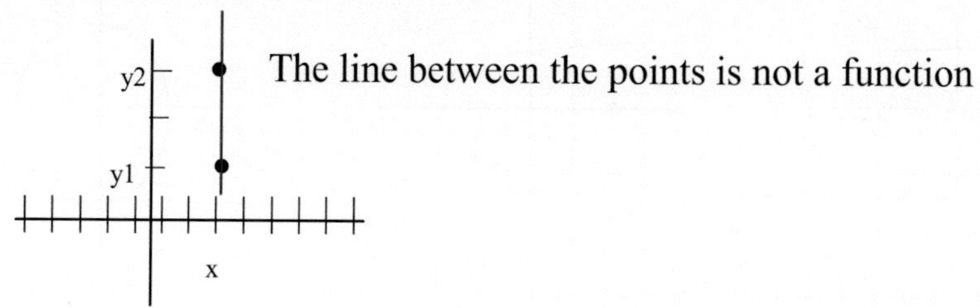

Example-2	Ordered Pairs as Relation

Determine whether the following relations are function:

A. {(1,3), (-4,7), (5,8), (1,11)}.
B. {(-2,4),(-2,7), (0,5), 4,8)}
C. {(-1,4), (-2,4), (-3,4), (-4,4)}
D. {(1,2), {3,7), {4,9), (5.11)}

Solution:

A. Yes
B. No
C. No
D. Yes

Example-3	Are the given equations Functions?

Determine which equation is a function?

A. $y=x^3$
B. $x=y^2$
C. $y=1/x$
D. $y = |x|$

Solution:

A. Yes
B. No
C. Yes
D. Yes

There are different types of functions as described in the table and their graphs:

Type of Functions
Polynomial Functions: $P(x)= ax^n$
Rational Functions: $R(x) = \frac{p(x)}{q(x)}$, $q(x) \neq 0$
Radical Functions: $F(x) = \sqrt{g(x)}$, $g(x) \geq 0$.
Power Functions: $H(x) = x^{n/m}$
Absolute Value Functions: $f(x) = \mid g(x) \mid$

Rational Radical Power Absolute value

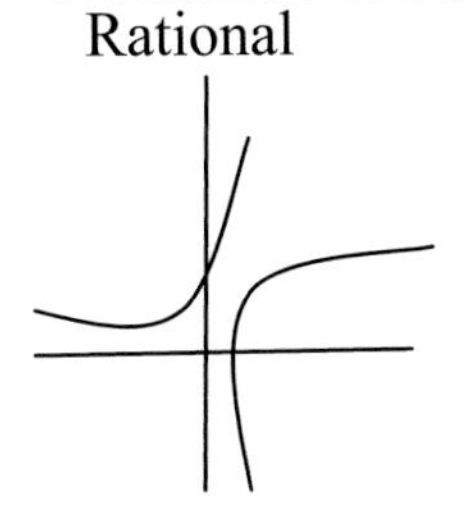

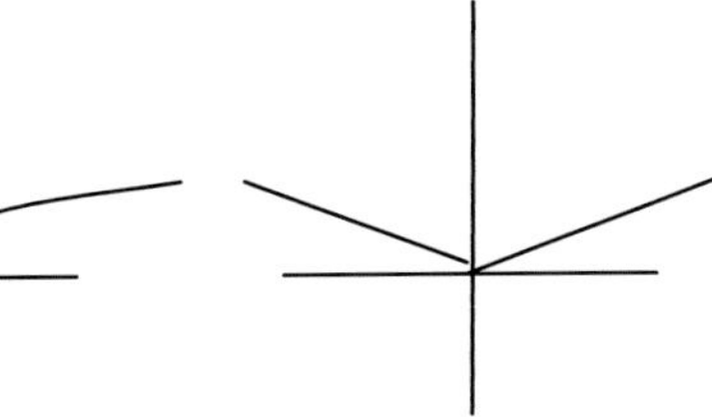

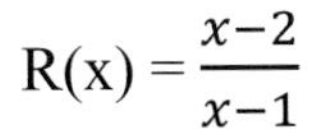

$R(x) = \frac{x-2}{x-1}$

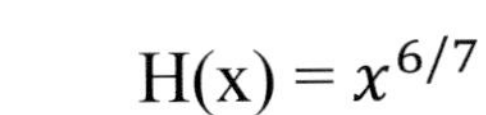

$F(x) = \sqrt{x-2}$

$H(x) = x^{6/7}$

$G(x) = \mid x-2 \mid$

Example-4	Determining if the Graph is a function

Determine whether the given graph is that of a function, if it is, use it to find to find its domain, range, intercepts, and symmetries:

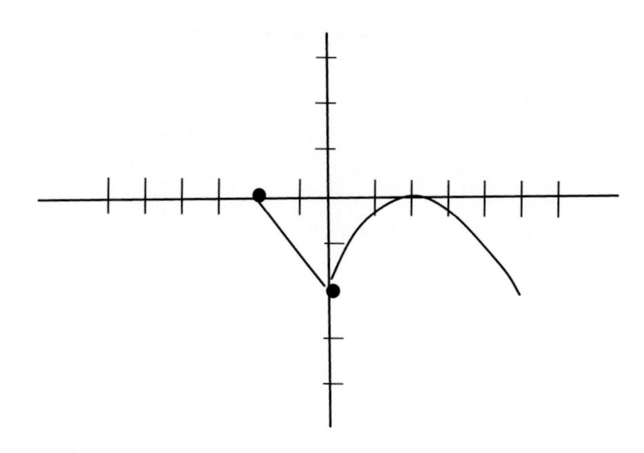

Solution:

* The graph is the graph of a function, because if we apply the vertical line test at any part of the graph the line will cross the function at one single point only.
* The domain of the function is: $\{x/\ x \geq -2\}$
* The range of the function is: $\{\ y/\ y \leq 0\}$.
* Intercepts are: (-2, 0), (0, -2), (2, 0)
* The graph is not symmetric about any axis.

Example-5	Finding the Domain of a Function

Find the domain of the following Functions:

A. $P(x) = x^3 + 3x^2 + 5$

B. $F(x) = \sqrt{x-7}$

C. $R(x) = \dfrac{x+1}{x-5}$

D. $H(x) = |\ x+3|$

Solution:

A. P(x) is a polynomial of degree n=3, the domain of polynomial is: all the real numbers $\{(-\infty,\infty)\}$

B. F(x) is a radical function, of even radical; to avoid non-real values we have to solve: $x-7 \geq 0$, which is: $x \geq 7$ then the domain of F(x) is: D:$\{\ x \geq 7\}$.

C. R(x) is a rational function; we have to avoid the undefined value by making the denominator Not equal to zero: $x-5 \neq 0 \rightarrow x \neq 5$.The domain of R(x) is D:$\{\ x \neq 5\}$

D. H(x) is absolute value function with domain D:{all the real numbers}.

1.2 Properties of Functions

A function is **even** if and only if its graph is symmetric with respect to the y-axis, and if for every number x in its domain, the number (–x) is also in the domain and f(–x) = f(x)
A function is **odd** if and only if its graph is symmetric with respect to the origin, and if for every number x in its domain, the number (–x) is also in the domain and f(-x) = –f(x).

Example-6 | Determining Even, Odd Functions from the Graph

Determine whether each given graph is Even, Odd, or Neither:

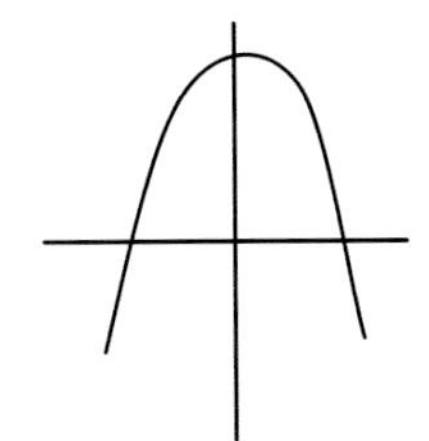

A. Even
Symmetric with respect to y-axis

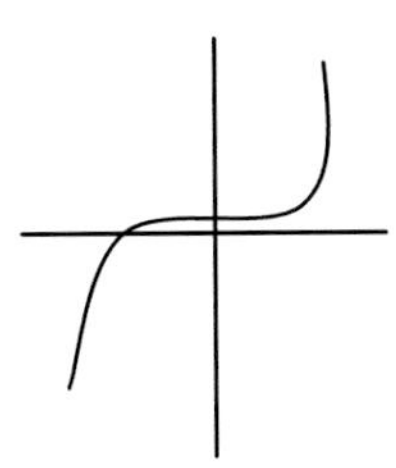

B. Odd
Symmetric with respect to the origin

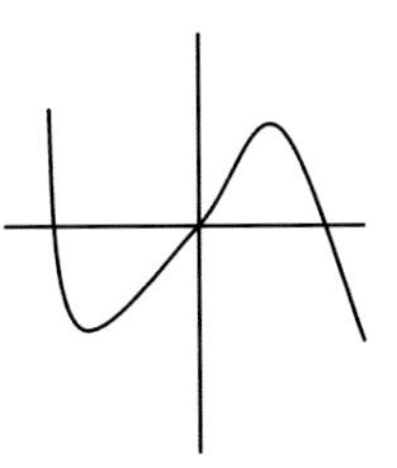

C. Neither
Not symmetric

Example-7 | Determining Even, Odd Functions Algebraically

Determine whether each of the given function is Even, Odd, or Neither:

A. $f(x) = x^4 + 5$
B. $f(x) = x^2 – x - 3$
C. $f(x) = x^3 + 2x$

Solution:
Replace each x with (–x):
A. $f(-x) = (-x)^4 + 5 = x^4 + 5 = f(x)$ → even function because it is symmetric with y-axis.
B. $f(-x) = (-x)^2 – (-x) – 3 = x^2 + x - 3 \neq f(x) \neq -f(x)$ → the function is neither even, nor Odd.
C. $f(-x) = (-x)^3 + 2(-x) = - x^3 - 2x = - (x^3 + 2x) = - f(x)$ → function is Odd, symmetric with respect to origin.

Example-8 | Determining Increasing, Decreasing Functions from the Graph

From the given graph, determine the intervals where the function is increasing, decreasing, or constant:

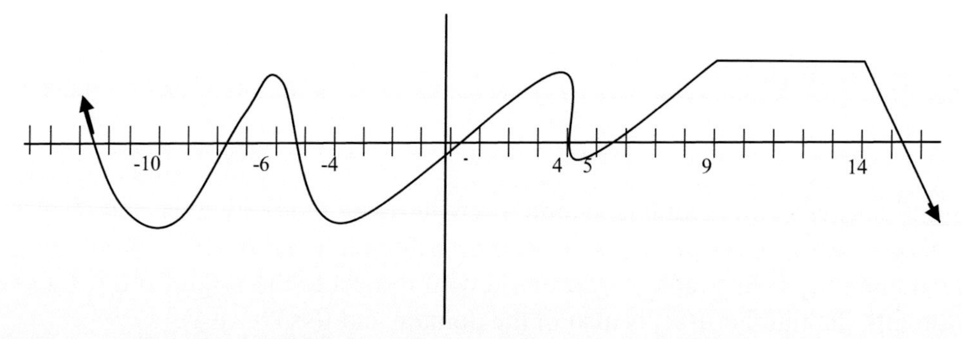

Solution: Function is Increasing on the following Intervals: (-10, -6), (-4, 4), (4.5, 9)
Function is decreasing on the following intervals: (-∞, -10), (-6, -4), (3.5, 4.5), (14,∞)
Function is constant on the interval: (9, 14)

Example-9	Determining Increasing, Decreasing Functions from the Graph

Use the graph to find the intervals on which it is increasing, decreasing or constant:

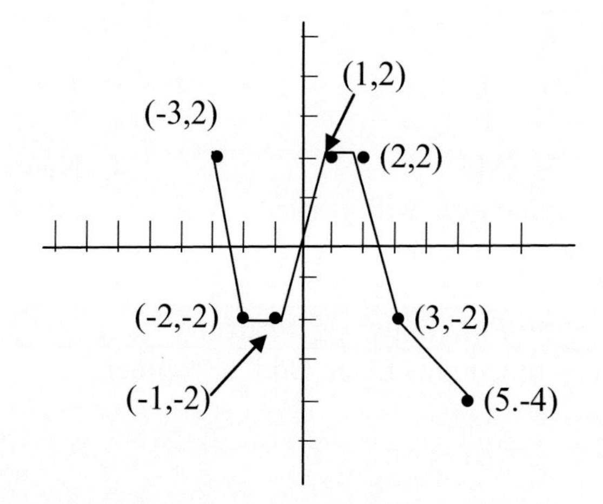

Solution:
* The graph is increasing on (–1, 1).
* The graph is decreasing on (–3,–2) and ((2, 5)
* The graph is constant on (–2,–1), and (1, 2)

1.3 The Average Rate of Change of a Function

Definition	The average Rate of Change of a Function y=f(x) from a to b, where a, b are in the domain, and a≠b is defined as: Average Rate of Change $= \frac{\Delta y}{\Delta x} = \frac{f(b)- f(a)}{b-a}$, a≠ b

Example-10	Finding the Average Rate of Change of the Function

Find the Average rate of change of $f(x) = 3x^3$:

A. From (–2) to 1

B. From 2 to 4

C. From (–1) to (– 4)

Solution: Using the above formula:

A. Average Rate of Change $= \frac{\Delta y}{\Delta x} = \frac{f(1) - f(-2)}{1-(-2)} = \frac{3-(-24)}{3} = \frac{27}{3} = 9$

B. Average Rate of Change $= \frac{\Delta y}{\Delta x} = \frac{f(4) - f(2)}{4-2} = \frac{192-24}{2} = \frac{168}{2} = 84$

C. Average Rate of Change $= \frac{\Delta y}{\Delta x} = \frac{f(-4) - f(-1)}{(-4)-(-1)} = \frac{-192-(-3)}{-3} = \frac{-189}{-3} = 63$

1.4 Piecewise Defined Functions

Example-11	Analyzing a Piecewise Defined Function

The function f is defined as:

$$F(x) = \begin{cases} x^2 + 1 & \text{if } x > 1 \\ 2 & \text{if } x = 1 \\ -x+1 & \text{if } -5 < x \le 1 \end{cases}$$

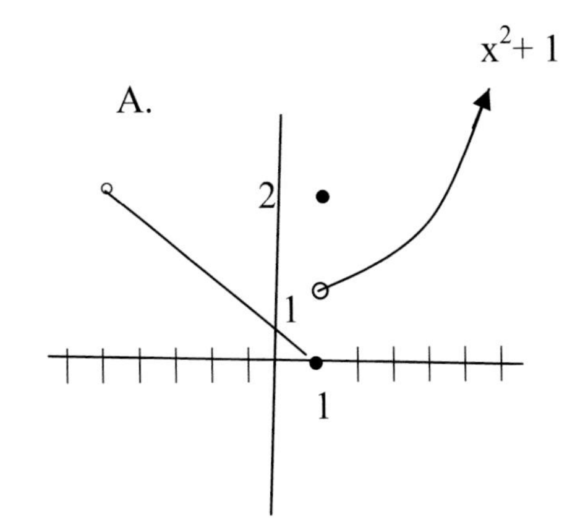

A. Graph the Function

B. Find F(-2), F(1), F(2)

C. Find the domain of F

D. Find the range of F

E. Find Intercepts

F. Is the function Continuous on its Domain?

Solution:

B. $F(-2) = -(-2) + 1 = 3$; $f(1) = 2$ (point); $F(2) = (2)^2 + 1 = 5$

C. Domain of F(x) is $(-5,\infty)$ or $x > -5$.

D. Range of F(x) is $y \ge 0$, or $[0,\infty)$

E. x-intercept: $f(x) = 0 \rightarrow x=1$; y-intercept: $F(0) \rightarrow y=1$

F. The function F is not continuous since there is a jump at x=1.

1.5 Transformation of Functions

Functions are transformed when they are:

Shifted to the Right → $y = f(x - h)$, $h > 0$

Shifted to the Left → $y = f(x+h)$, $h > 0$

Shifted Up → $y = f(x) + k, k > 0$
Shifted Down → $y = f(x) - k, k > 0$
Rotated about the x-axis → $y = - f(x)$
Rotated about the y-axis → $y = f(-x)$
Stretched vertically → $y = a f(x), a > 1$
Compressed Vertically → $y = a f(x), 0 < x < 1$.
Stretched horizontally → $y = f(ax), 0 < x < 1$
Compressed horizontally → $y = f(ax), a>1$.
For all these transformation one formula can be applied:
$y = \pm a f(bx \pm h) \pm k$
We will apply all these transformation to the function $f(x) = x^2$, first using table then graph.

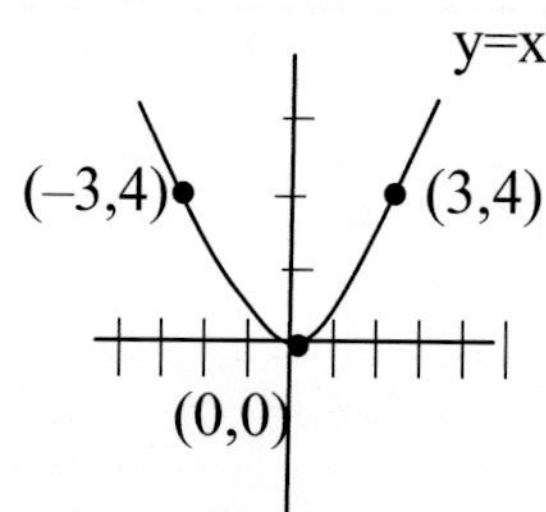

Transform the given function $y=x^2$ as follows:
A. Shift $y= x^2$ to the right by 2-units.
B. Shift $y= x^2$ to the left by 2-units.
C. Shift $y= x^2$ up by 2-units.
D. Shift $y= x^2$ down by 2-units.
E. Rotate $y= x^2$ about the x-axis
F. Rotate $y= x^2$ about the y-axis
G. Stretch $y= x^2$ vertically by 2-units
H. Compress $y= x^2$ vertically by 2-units
I. Stretch $y= x^2$ horizontally by 2-units
J. Compress $y= x^2$ horizontally by 2-units.

$y=x^2$ (x,y)	A (x+2, y)	B (x-2, y)	C (x, y+2)	D (x, y-2)	E (x,- y)	F (-x, y)	G (x, 2y)	H (x,1/2y)	I (2x,y)	J (1/2x, y)
(-3, 4)	(-1, 4)	(-5,4)	(-3,6)	(-3, 2)	(-3, -4)	(3,4)	(-3, 8)	(-3,2)	(-6,4)	(-3/2,4)
(0,0)	(2,0)	(-2,0)	(0, 2)	(0,-2)	(0,0)	(0,0)	(0,0)	(0,0)	(0,0)	(0,0)
(3,4)	(5,4)	(1,4)	(3,6)	(3, 1)	(3, -4)	(-3,4)	(3,8)	(3,2)	(6,4)	(3/2,4)

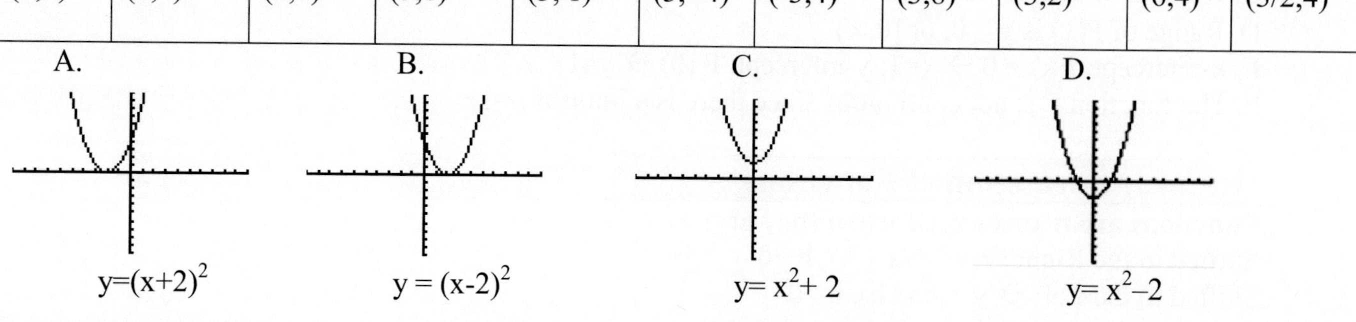

E.

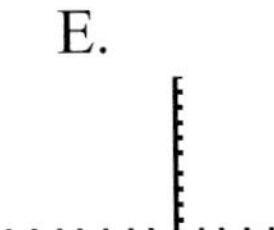

$y= -x^2$

F.

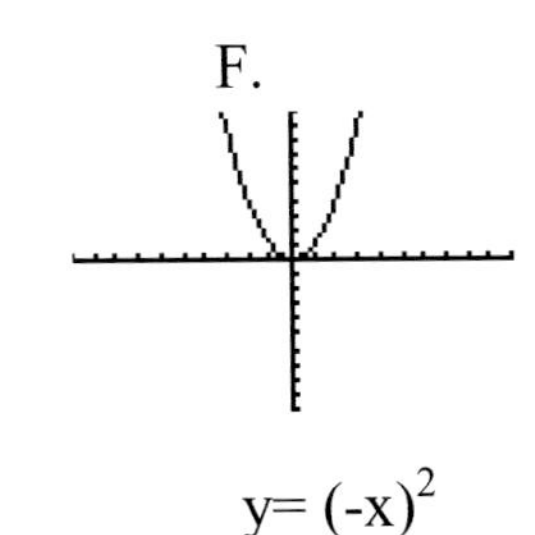

$y= (-x)^2$

G.

$y= 2x^2$

H.

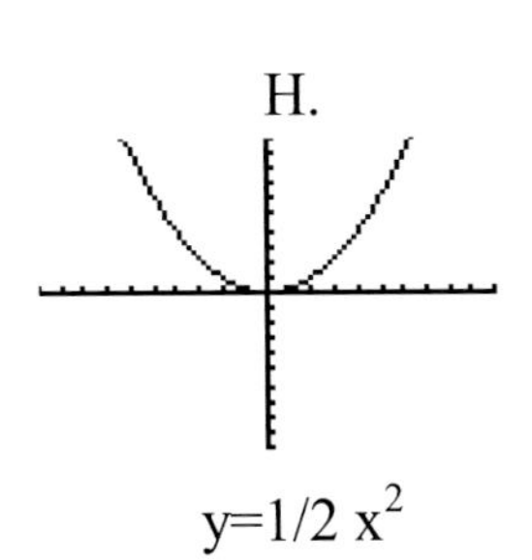

$y=1/2\ x^2$

Example-12	**Graph a function using transformation**

Graph the polynomial using transformation method: $p(x) = (x+1)^3 - 4$

Solution: 1. The original polynomial without transformation is $p(x) = x^3$.

2. It was shifted to the left by 1-unit $p(x) = (x+1)^3$.

3. It was shifted down by 4-units $p(x) = (x+1)^3 - 4$ as shown in the following graphs:

1.

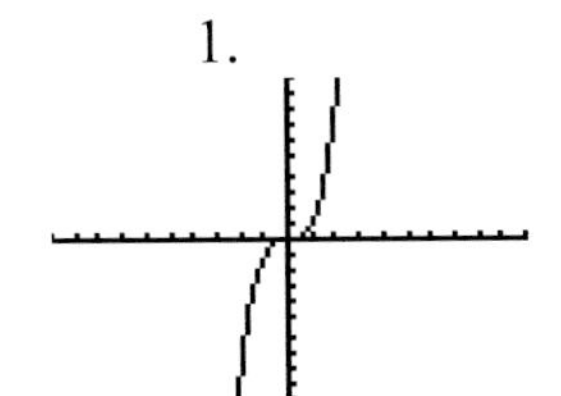

2.

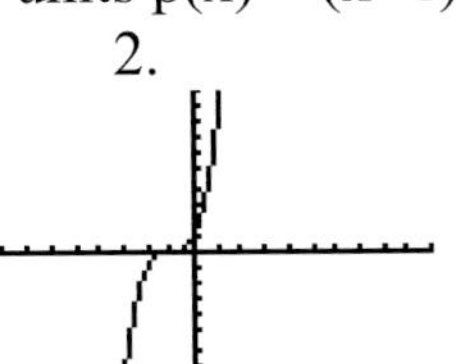

3.

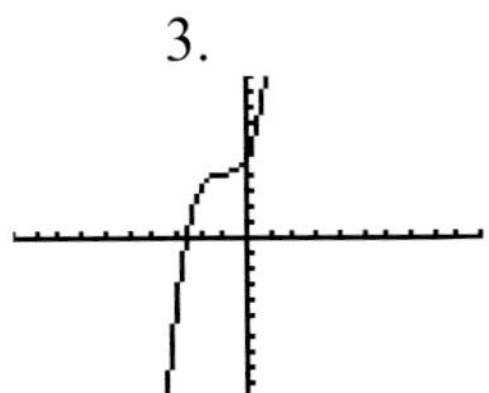

Definition	**Vertical Stretching and Shrinking** For the graph of the function y=f(x) transformed as y = af(x) will be: * Stretched vertically if $\lvert a\rvert >1$, and * Compressed vertically if $0< \lvert a\rvert < 1$. **Horizontal Stretching and Shrinking** For the graph of the function y=f(x) transformed as y = f(ax) will be: * Stretched horizontally if $\lvert a\rvert >1$, and * Compressed horizontally if $0< \lvert a\rvert < 1$.

Example-13	**Horizontal and Vertical, Shrinking and Stretching**

Show the vertical and Horizontal Stretching and Shrinking on the given graph y=f(x):

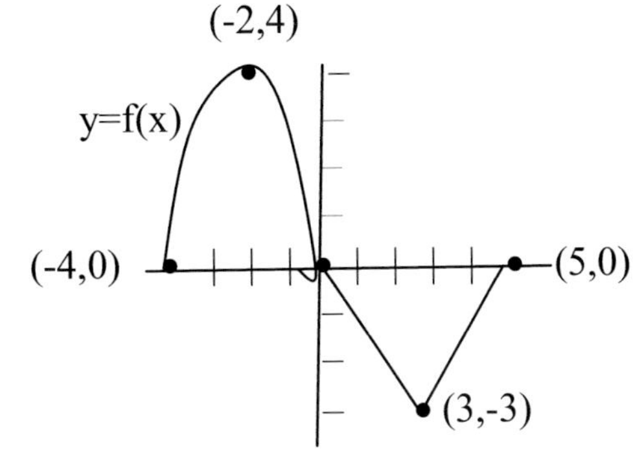

A. y=2f(x)
B. y=1/2f(x)
C. y=f(2x)
D. y=f(1/2x)

Solution:

y=f(x) (x,y)	A. y=2f(x) (x,2y)	B. y=1/2 f(x) (x,1/2y)	C. y=f(2x) (2x,y)	D. y=f(1/2x) (1/2x, y)
(-4,0)	(-4,0)	(-4,0)	(-8,0)	(-2,0)
(-2,4)	(-2, 8)	(-2,2)	(-4,4)	(-1,4)
(0,0)	(0,0)	(0,0)	(0,0)	(0,0)
(3,-3)	(3,-6)	(3, -3/2)	(6,-3)	(3/2, -3)
(5,0)	(5,0)	(5,0)	(10,0)	(5/2,0)

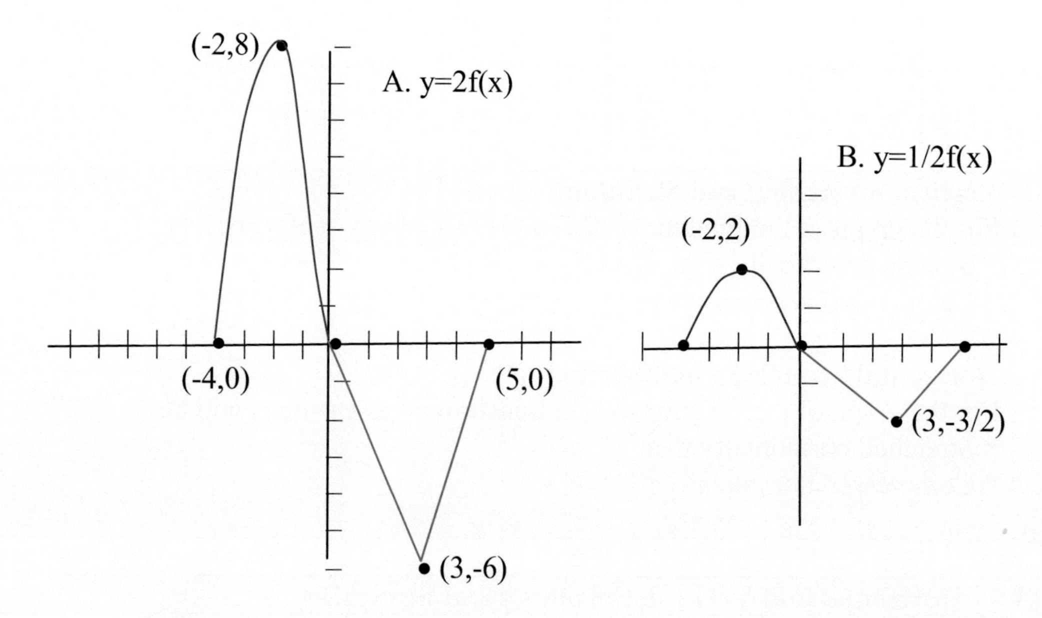

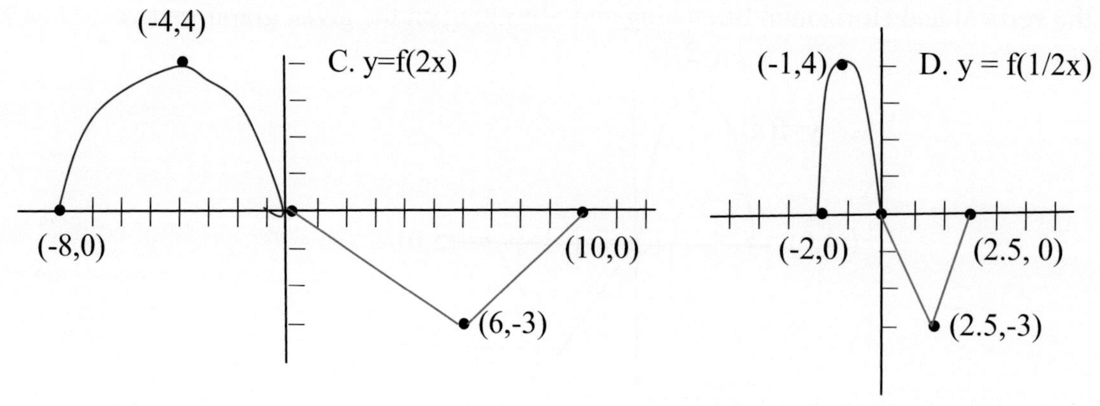

Example-14	Finding the domain of a piecewise function

For the given function f(x):

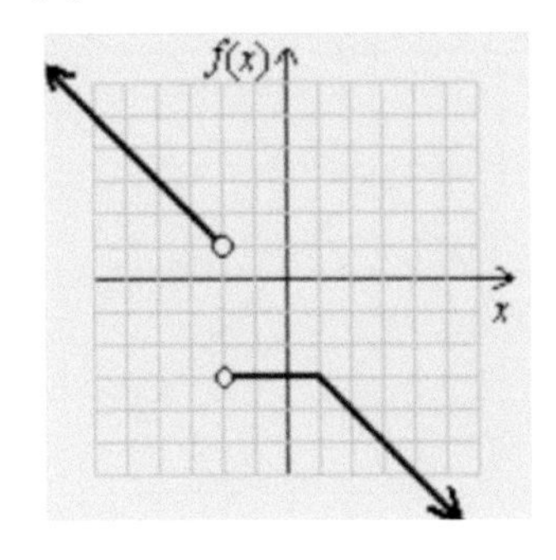

A. Find the domain
B. Find the range
C. Find the x-intercept(s)
D. Find the y-intercept(s)
E. Find the interval over which f(x) is increasing
F. Find the interval over which f(x) is decreasing
G. Find the intervals over which f(x) is constant
H. Find any pints of discontinuity.

Solution:
A. Domain of f(x) is $(-\infty, -2) \cup (-2, \infty)$
B. Range of f(x) is $(-\infty, -3] \cup (1, \infty)$
C. None
D. -3
E. None
F. $(-\infty, -2)$, $[1, \infty)$
G. $(-2, 1]$
H. $x = -2$

Example-15	Real Zeros of Polynomial

From the graph:
A. List all the real zeros of the polynomial
B. List the turning points of p(x)
C. State the left and right behavior of the function p(x)

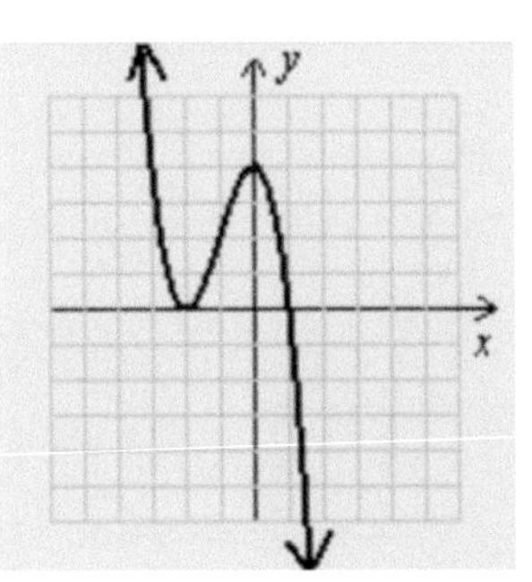

Solution:
A. Real zeros of p(x) at : -2,1
B. Turning points at: (-2,0),(0,4)
C. $P(x) \to -\infty$ as $x \to \infty$ and $P(x) \to \infty$ as $x \to -\infty$

Example-16	Is the graph of a polynomial function?

State if the graph is of a polynomial function. If not, state why?

Solution: The graph is not a polynomial function, because it has a sharp edge.

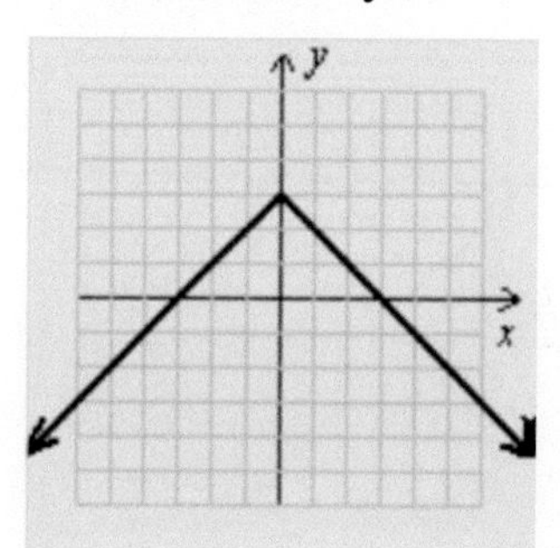

Example-17	Graphing a polynomial function

Graph the polynomial function $p(x) = x^4 + 3x^3 - 3x^2 - 11x - 6$, and state its real zeros.

Solution:

The polynomial is of degree n = 4, then the number of real zeros is expected to be 4,as shown in the graph: zeros at: -3 of multiplicity 1

-1 of multiplicity 2

2 of multiplicity 1

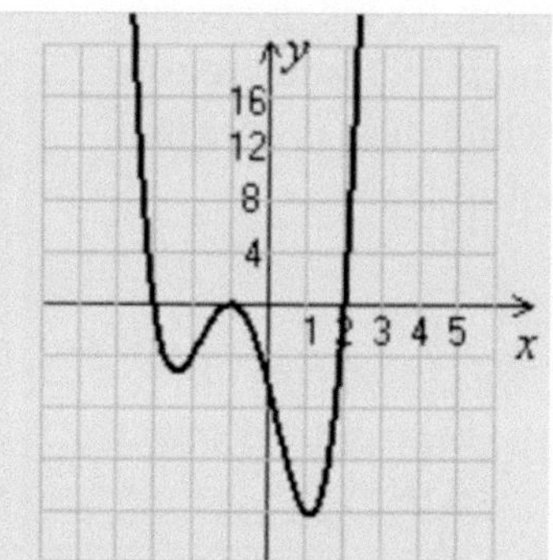

Example-18	Polynomial Inequality

Solve the inequalities:

A. $P(x) \geq 0$ using the graph of $P(x)$ given in A. below.

B. $P(x) < 0$ using the graph of $P(x)$ given in B. below.

Solutions: The solution is:

A. $(-\infty, -3] \cup [-1, 4]$ B. $(-3, -1) \cup (4, \infty)$

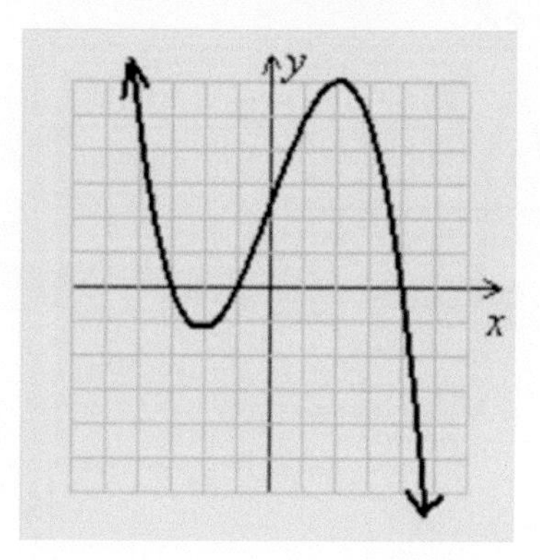

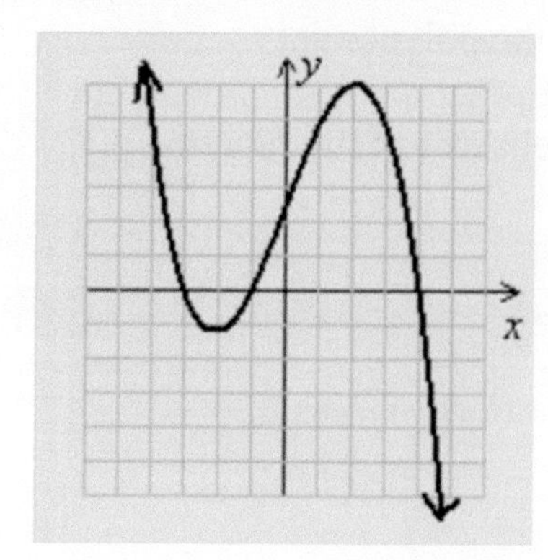

Chapter - 1 Exercise

1. Determine if the relation represent a function. If it's a function, state the domain and range:

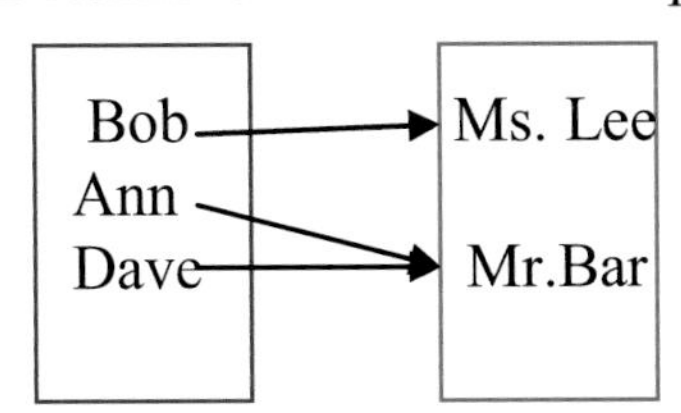

2. Determine if the relation represent a function. If it's a function, state the domain and Range :{(29,-2), (4,-1), (4,0), (5,1), (13,3)}
3. Determine if the function is one-to-one function:

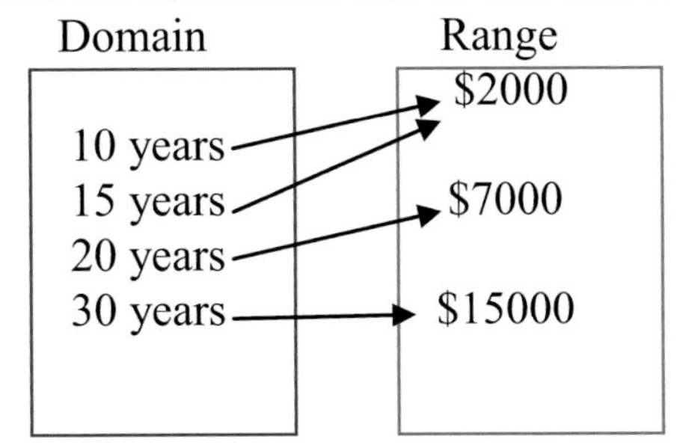

4. Determine whether the function is one-to-one.
   {(-3,-2), (-2,-2), (-1,8), (0,-5)}
5. Find the inverse of the function and state its domain and range:
   {(-3,4), (-1,5), (0,2), (2,4), (5,7)}
6. The function f(x) is one-to-one function. Find its inverse:

$$f(x) = \frac{3x+1}{2}$$

7. The function f(x) is one-to-one function. Find its inverse:
   $F(x) = (x+2)^3 - 8$

8. Find the function that is finally graphed after the following transformations are applied to the Graph of y= √ x. The graph is shifter down 6-units, reflected about the y-axis, and finally Shifted right 2-uunits.

9. Determine the domain and range of the function: $f(x) = x^2 - 6x + 8$.
10. Solve the inequality: $x^4 - 13x^2 < -36$
11. Solve the inequality: $30 - 13x - 2x^2 + 4x^3 \geq x^4$
12. Find the degrees of the polynomial: $p(x) = (x+6)(x-8)^3(x-1)^2$

2. Polynomial Functions

Chapter – 2
Polynomial Functions

Objectives: 2.1 Polynomial functions and Graphs
2.2 Zero's of polynomials
2.3 Polynomial Division, The Remainder Theorem
2.4 Polynomial Inequalities
2.5 Applications

2.1 Polynomial Functions and Graphs

Polynomial Functions	Polynomial functions can be described in the standard form as: $P(x) = a_n x^n + a_{n-1} x^{n-1} + \ldots\ldots + a_1 x^1 + a_0$ …………… (1) Where: n is the degree of the polynomial a_n is the leading coefficient of the polynomial a_0 is the constant of the polynomial
Rules of Polynomials	Multiplication → $x^n . x^m = x^{n+m}$ Division → $x^n / x^m = x^{n-m}$ Power → $(x^n)^m = x^{nm}$ $x^{-n} = 1/x^n$ $1/x^{-n} = x^n$
Parts of Polynomials	One term of the polynomial is called Monomial Two terms of the polynomial is called Binomial. Three terms of the polynomial is called Trinomial..etc

Polynomial is any function that is written in the form of (1).
Functions are not just polynomials; there are many other functions with different names.
Rewriting the polynomial function (1) in the ascending order as:

$$\mathbf{P(x) = a_0 + a_1x^1 + a_2 x^2 + a_3 x^3 + \ldots\ldots + a_{n-1} x^{n-1} + a_n x^n \ldots\ldots\ldots\ldots\ldots \quad (2)}$$

Will help us to split it in ascending order degrees as follows:

1) $P(x) = a_0$ is a polynomial of degree n=0 (constant).
The graph of the constant is a horizontal line:

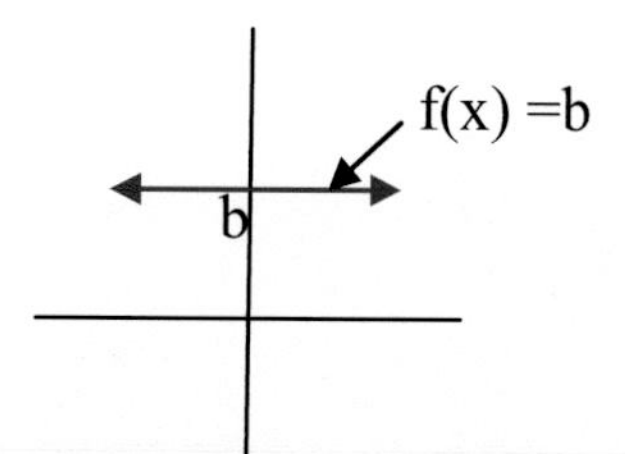

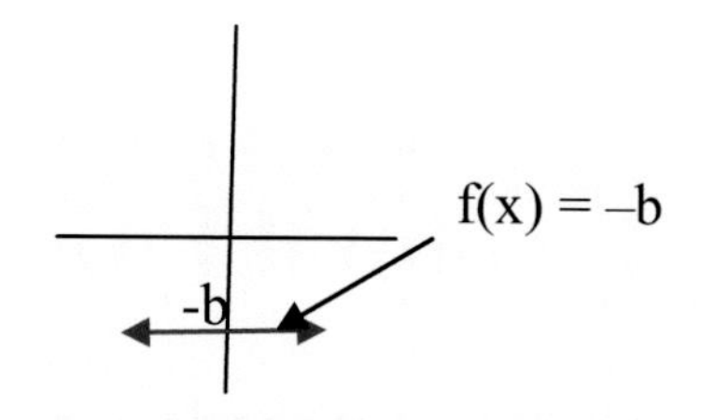

Analysis of Constant function
Domain: All the real numbers
Range: ± b
x-intercept: None
y-intercept: ± b

2) $P(x) = a_0 + a_1x^1$ is a polynomial of degree n=1 (linear).
The graph of linear polynomial is a straight line with slope (± m):

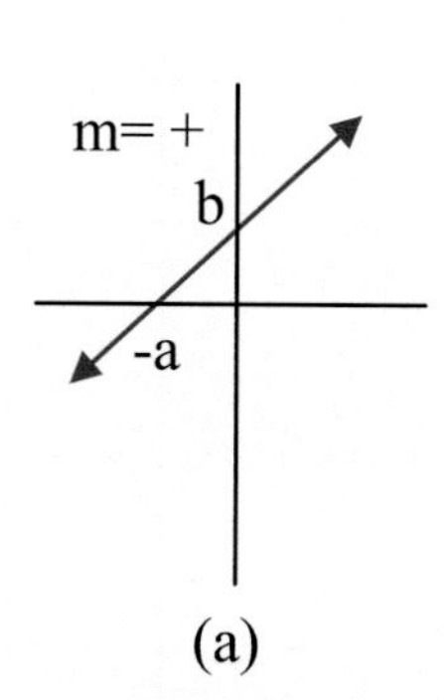

(a)

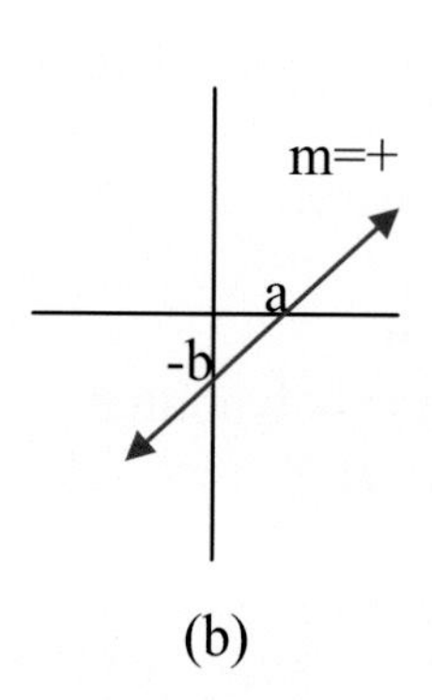

(b)

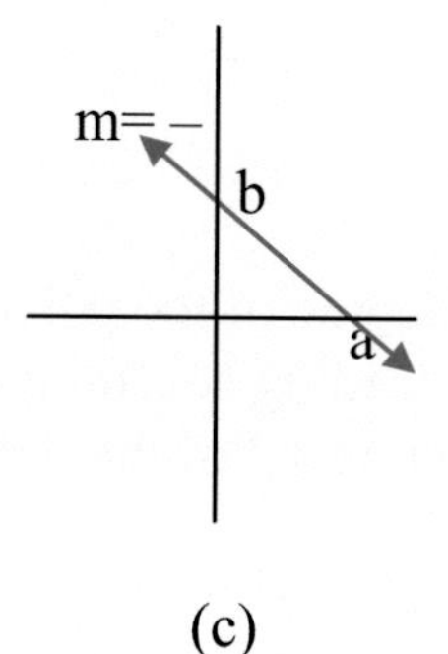

(c)

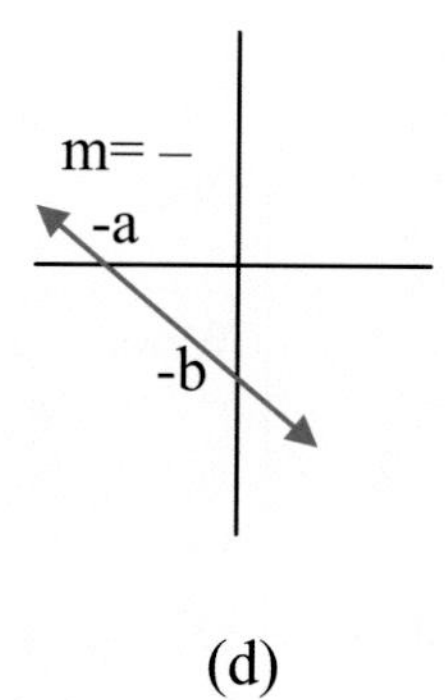

(d)

Analysis of linear functions
Domain: All the real numbers {-∞,∞}
Range: All the real numbers {-∞,∞}
x-intercepts: ± a
y-intercepts: ± b

3) $P(x) = a_0 + a_1x^1 + a_2 x^2$ is a polynomial of degree n=2 (quadratic)
The graph of quadratic is a parabola, opened up, or opened down depending on the leading Coefficient a_2 whether it's positive or negative:

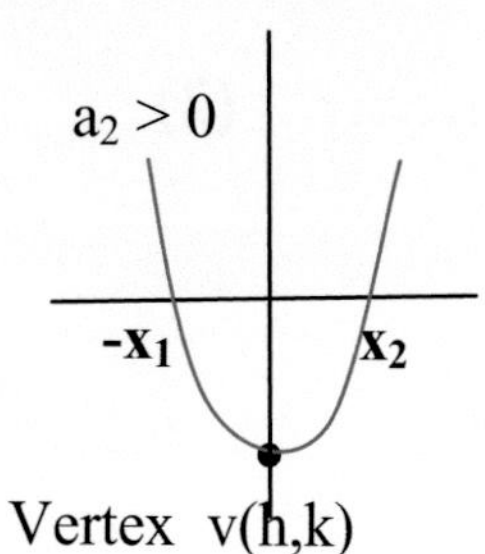

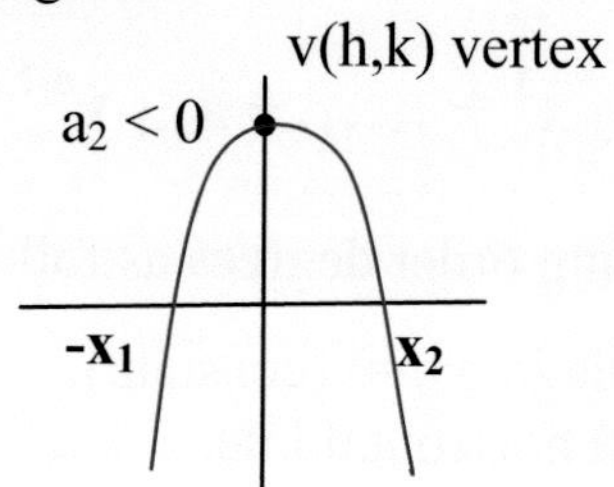

Analysis of quadratic functions
Domain: All the real numbers $\{-\infty,\infty\}$
Range: (k,∞) if opened up → $a_2 > 0$
$(-\infty, k)$ if opened down → $a_2 < 0$
x-intercepts: $\{x_1 , x_2\}$
y-intercepts: $\pm$ y

4) $P(x) = a_0 + a_1 x^1 + a_2 x^2 + a_3 x^3$ is a polynomial of degree n=3 (cubic)

The graph of cubic function is opened up or down depending on the leading coefficient a_3:

$a_3 > 0$

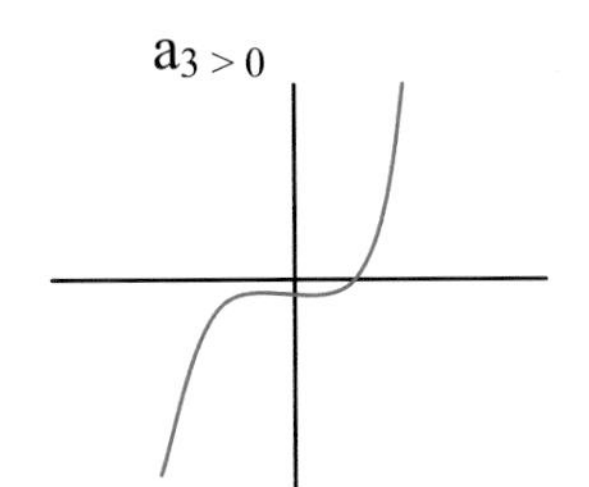

$a_3 < 0$

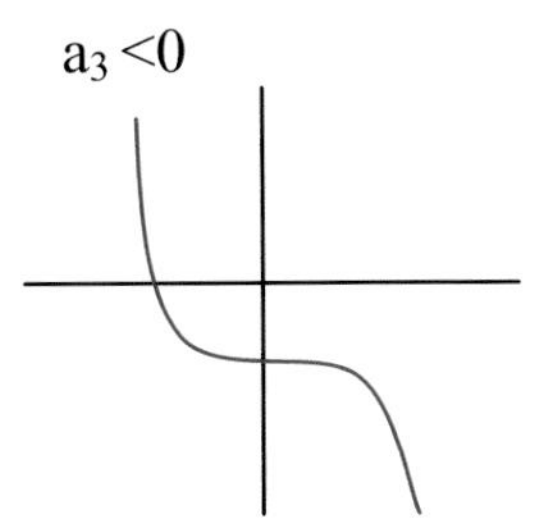

Analysis of Cubic functions
Domain: All the real numbers $\{-\infty,\infty\}$
Range: All real numbers $(-\infty,\infty)$
x-intercepts: $\pm$ x
y-intercepts: $\pm$ y

Example - 1 | Matching functions to graphs

Match each of the following graphs to the given functions, by using the leading term:

a) $f(x) = 3x^2 + 2x - 3$
b) $f(x) = -3x^3 - x^2 + 4x - 1$
c) $f(x) = 4x^4 + 2x^3 + 5$
d) $f(x) = - x^5 - 4 x^3 + 6$

A

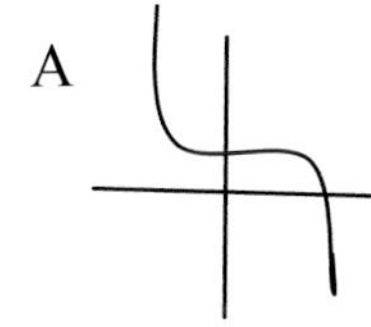

B

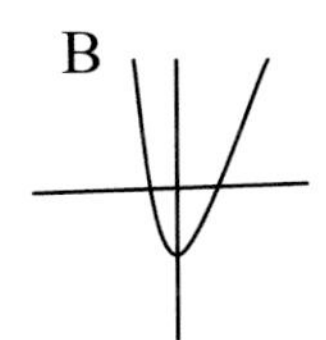

C

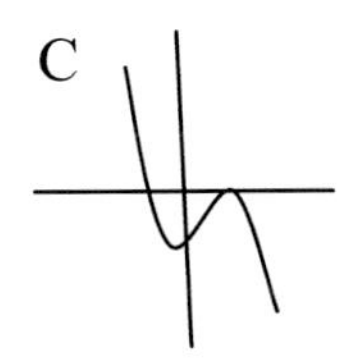

D

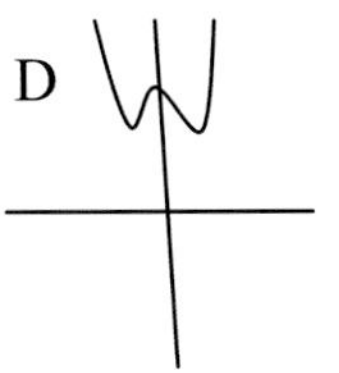

Solution:

a) The degree is n=2 and a > 0 → the matching graph is B.
b) The degree is n=3 and a < 0 → the matching graph is C.
c) The degree is n=4 and a > 0 → the matching graph is D.
d) The degree is n=5 and a < 0 → the matching graph is A.

Polynomial degrees and Leading Coefficient

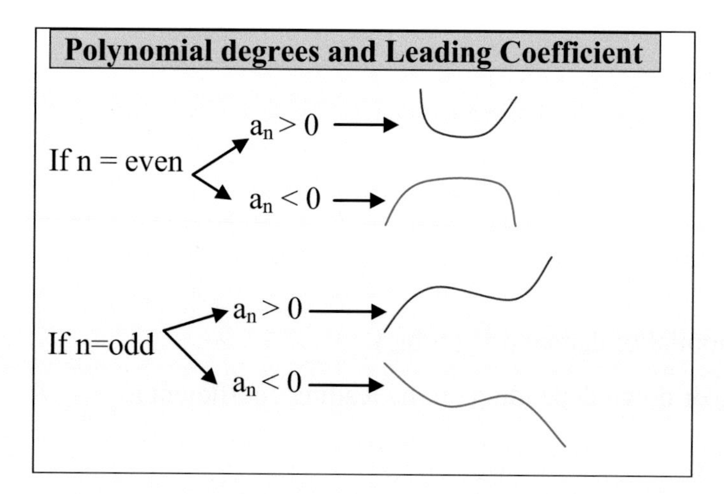

2.2 Zero's of Polynomial Functions

Finding the zero's of polynomial of degree n = 2

Polynomial of degree n=2 is called the quadratic functions.
The quadratic functions has a shape of a parabola, opened up or down depending on the sign of the leading coefficient if positive, then it will open up, if negative then it will open down.
The solution can be found by the intersection of the parabola with x-axis or the intercepts, if the parabola does not intersect the x-axis that is when the solution is not real, but complex, as shown on the given graphs:

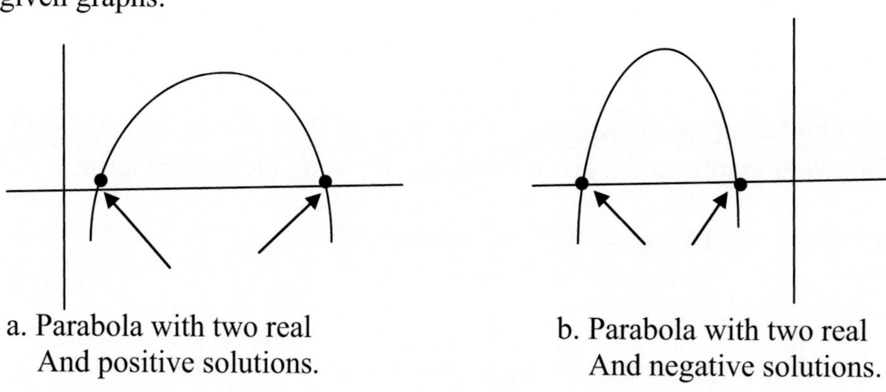

a. Parabola with two real And positive solutions.

b. Parabola with two real And negative solutions.

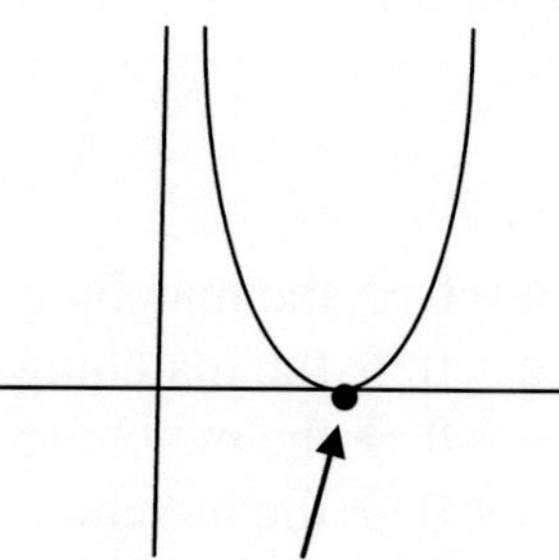

c. Parabola not crossing the x-axis Means complex solution

d. Parabola touching the x-axis means double solutions.

Example-2	Finding the zero's of Quadratic Polynomial n = 2

Find the zero's of the function: $p(x) = (x+1)(x-5)$

Solution: To find the zero's of p(x), we let $p(x) = 0$ and solve: $(x+1)(x-5) = 0$ for x:

$x+1 = 0 \rightarrow$ or $x=-1$ and $x-5 = 0 \rightarrow$ or $x=5 \rightarrow$ they are the zero's of p(x).

That is: $p(-1) = 0$, and $p(5) = 0$

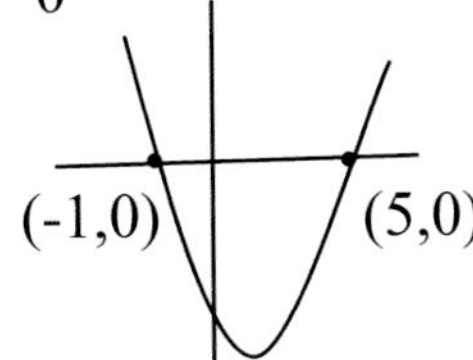

Example-3	Finding the zero's of Cubic Polynomial n=3

Find the zero's of the function: $p(x) = (x+1)(x-2)(x+3)$

Solution: Let $p(x) = 0$ and solve $(x+1)(x-2)(x+3) = 0$, we get:

$x=-1$, $x=2$, and $x=-3 \rightarrow$ they are the zero's of the cubic function p(x).

That is: $p(-1) = 0$, $p(2) = 0$, and $p(-3) = 0$.

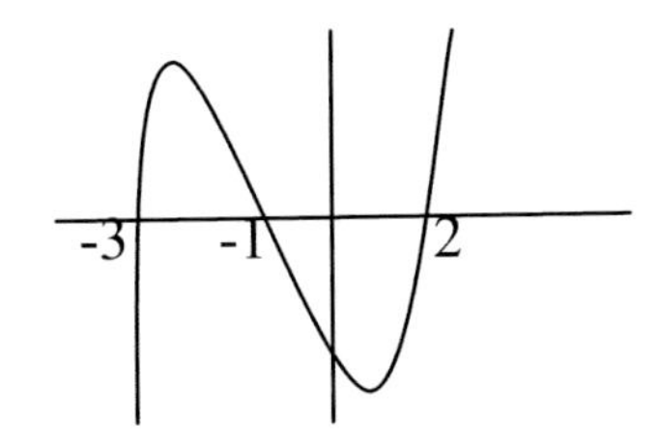

Multiplicity of Zero's of Polynomial

If $(x-c)^n$ is a factor of a polynomial, if:

* n is even → then the graph touches the x-axis at (c,0)
* n is odd → then the graph crosses the x-axis

Example-4	Finding the Multiplicity zero's of Polynomial

Find the zeros of $p(x) = (x-1)(x-1)(x+3)$

Solution: The zeros are: $x=1$ of multiplicity 2 (even) → the graph touches the x-axis at 1, and $x=-3$ of Multiplicity 1 → the graph crosses the x-axis at -3.

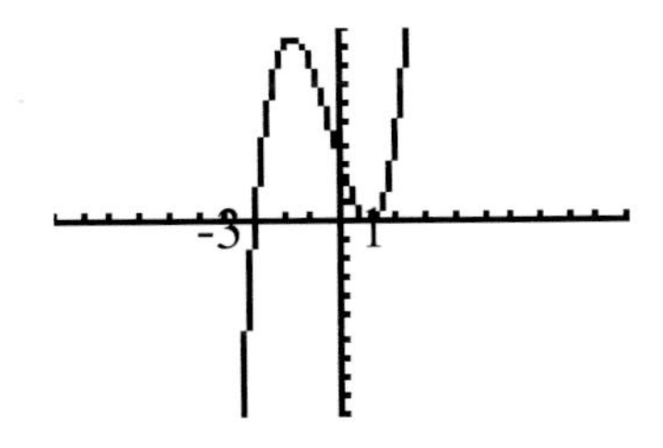

Example-5	Finding the Multiplicity zero's of Polynomial

Find the zero's of $p(x) = -(x+1)(x+1)(x-2)(x-2)$

Solution: Since $a< 0$ then p(x) is opened down, and the zero's are: x=-1 of multiplicity-2 (even) then the graph touches the x-axis at x=-1, and x=2 of multiplicity-2(even) which means the graph touches the x-axis at x=2.And the polynomial is of degree n=4.

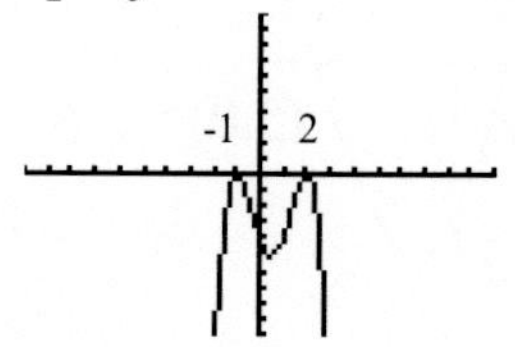

The Intermediate Value Theorem

A polynomial function p(x) with p(a), p(b), of opposite sign, where $a\neq b$ → then the function has a real zero between a and b.

P(a)
(a,p(a))
P(c) = 0
P(b)
(b,p(b))

Example-6	Using Intermediate Value Theorem

Determine whether the function has a real zero's between a, and b:

A. $p(x) = x^3 + 2x^2 - 3x$; $a = -1$, $b = -4$

B. $h(x) = x^4 - 2x^2 - 6$; $a = 2$, $b = 3$

C. $g(x) = 2x^5 - 7x + 1$; $a = 1$, $b = 2$

Solution: **A.** $p(-1) = (-1)^3 + 2(-1)^2 - 3(-1) = 4$ (positive)

$P(-4) = (-4)^3 + 2(-4)^2 - 3(-4) = -20$(negative)

Then we expect a zero between (–1), and (–4).

B. $h(2) = (2)^4 - 2(2)^2 - 6 = 2$ (positive)

$h(3) = (3)^4 - 2(3)^2 - 6 = 57$ (positive)

We don't expect a zero between 2 and 3.

C. $g(1) = -4$ (negative); $G(2) = 51$ (positive) → we expect a zero between 1, and 2.

2.3 Polynomial Division, the Remainder Theorem

Objectives: 1. Long Division
2. Synthetic Division

1. Long division: terms are used as they are with variables and coefficients, and powers in descending order, if one order term is missing it should be replaced with a zero, or empty space.

Division Algorithm for two polynomials p(x) , g(x)

$$\frac{P(x)}{g(x)} = Q(x) + \frac{R(x)}{g(x)}$$ → or $p(x) = g(x) . Q(x) + R(x)$ ...... (1)

p(x) → dividend, g(x) → divisor, Q(x) → quotient, R(x) → remainder

Example -7	Divide the polynomial by binomial: $3x^3 + x^2 + 4x - 6$ by $x + 3$

Solution: to divide $\frac{3x^3 + x^2 + 4x - 6}{x + 3}$ we follow these steps:

$$\begin{array}{r|l} & 3x^2 - 8x + 28 \\ \hline x+3 & 3x^3 + x^2 + 4x - 6 \\ & +3x^3 + 9x^2 \quad \leftarrow \text{Change the sign of this raw} \\ \hline & -8x^2 + 4x \\ & -8x^2 - 24x \quad \leftarrow \text{Change the sign of this raw} \\ \hline & +28x - 6 \\ & +28x + 84 \quad \leftarrow \text{Change the sign of this raw} \\ \hline & -90 \quad \leftarrow \text{this is the remainder} \end{array}$$

Then $\frac{3x^3 + x^2 + 4x - 6}{x + 3} = (3x^2 - 8x + 28) + \frac{-90}{x + 3}$

Dividend: $3x^3 + x^2 + 4x - 6$; Quotient: $3x^2 - 8x + 28$; Remainder: -90; Divisor: $x + 3$

Practice -7	Divide the polynomial by binomial: $2x^4 - x^3 + 16x^2 - 4$ by $2x - 1$

2. Synthetic Division: This method is a short cut for long division, it concentrates on Coefficients only, as before if a power term is missing it should be replaced with a Space, a zero, or dashed area.

Example -8	Divide the polynomial by binomial: $2x^3 + 4x^2 + 8x - 8$ by $x + 2$

Solution: to divide $\dfrac{3x^3 + 4x^2 + 8x - 8}{x + 2}$ we follow these steps:

$$\begin{array}{r|rrrr} -2 & 3 & 4 & 8 & -8 \\ & & -6 & 4 & -24 \\ \hline & 3 & -2 & 12 & -32 \end{array} \leftarrow \text{Remainder}$$

Then $\dfrac{3x^3 + 4x^2 + 8x - 8}{x + 2} = (3x^2 - 2x + 12) + \dfrac{-32}{x + 2}$

Practice -8	**Divide the polynomial by binomial: $2x^5 - 2x^4 + 3x^2 - x + 1$ by $x - 2$**

Remainder Theorem

If the divisor $g(x) = x - c$, where c is a real number, then (1) becomes:
$p(x) = (x - c)\,Q(x) + R$ .... (2)
and if $x = c \rightarrow$ then $p(x) = p(c) = R$, then (2) can also be written as:
$p(c) = (x - c)\,Q(c) + R$ .... (3)

Example-9	Synthetic Division and Remainder Theorem

Given that $p(x) = 2x^3 + 7x^2 - 5$. Find p (–3) using both synthetic division, and the remainder theorem

$$\begin{array}{r|rrrr} -3 & 2 & 7 & 0 & -5 \\ & & -6 & -3 & 9 \\ \hline & 2 & 1 & -3 & 4 \leftarrow \text{Remainder} \end{array}$$

Using remainder theorem → $p(-3) = 2(-3)^3 + 7(-3)^2 - 5 = 4$

Finding polynomial with given zero's

Every polynomial function p(x) with degree n, $n \geq 1$, can be factored into n-linear factors:

$$\mathbf{P(x) = a_n (x - c_1)(x - c_2) \ldots (x - c_n) \ldots\ldots\ldots (4)}$$

Number of Real Zero's

A polynomial function of degree n has n-zeros.

Example-10	Finding a Polynomial function from given Real Zero's

Find the polynomial function of degree n= 3, and zeros: -1, 0, 4.

Solution: Using the above theorem, where given: $c_1 = -1, c_2 = 0, c_3 = 4$, then the polynomial is:

$$\begin{aligned} P(x) &= a_n (x + 1)(x - 0)(x - 4) \\ &= a_n x (x + 1)(x - 4) \\ &= a_n x (x^2 - 3x - 4) \\ &= a_n (x^3 - 3x^2 - 4x), \text{ letting } a_n = 1 \text{ gives:} \end{aligned}$$

→ $P(x) = x^3 - 3x^2 - 4x$

The fundamental Theorem of Algebra / Complex Zero's

A polynomial function p(x) of degree n, $n \geq 1$, has at least one set of non real (complex) number→ $a \pm bi, b \neq 0$

Example-11	Finding a Polynomial function from given Complex Zero's

Find the polynomial p(x) of degree n=4, and given zero's: 1, 1, -2+i.

Solution:

Using formula (5) with $a_n = 1$ and $c_1 = 1$, $c_2 = 1$, $c_3 = -2+I$, and $c_4 = -2-i$:

$$\begin{aligned} P(x) &= (x-1)(x-1)[x-(-2+i)][x-(-2-i)] \\ &= (x^2 - 2x + 1)[x^2 - x(-2-i) - x(-2+i) + (-2-i)(-2+i)] \\ &= (x^2 - 2x + 1)[x^2 + 2x + xi + 2x - xi + (4 - i^2)] \\ &= (x^2 - 2x + 1)[x^2 + 4x + (4 - (-1))] \end{aligned}$$

$= (x^2 - 2x + 1)(x^2 + 4x + 5)$
$= x^2(x^2 + 4x + 5) - 2x(x^2 + 4x + 5) + 1(x^2 + 4x + 5)$
$= x^4 + 4x^3 + 5x^2 - 2x^3 - 8x^2 - 10x + x^2 + 4x + 5$

Then → $P(x) = x^4 + 2x^3 - 2x^2 - 6x + 5$

Descartes' Rule of Signs

* The number of positive real zero's of p(x) = the number of variation Of sign in p(x) or less by a positive even integer.
* the number of negative real zero's = the number of variation sign in p(−x), or less by a positive even integer

Potential Zero's

For polynomial of degree n:

$P(x) = a_n x^n + a_{n-1} x^{n-1} + \ldots\ldots + a_1 x^1 + a_0$ .... (4)

$$\text{The potential zero's are} = \frac{a_0}{a_n} = \frac{\text{constant of } p(x)}{\text{leading coefficient of } p(x)}$$

Example -12 | Factors of Polynomial Functions

For the given polynomial: $p(x) = 3x^4 + 5x^3 + 25x^2 + 45x - 18$

A. Use Descartes 'Rule to find the positive and negative zeros.
B. Write the potential zeros.
C. Use synthetic division to find the factors

Solution:

A. $p(x) = 3x^4 + 5x^3 + 25x^2 + 45x - 18$

1-variation sign → 1-positive zero

$p(-x) = 3(-x)^4 + 5(-x)^3 + 25(-x)^2 + 45(-x) - 18$
$= 3x^4 - 5x^3 + 25x^2 - 45x - 18$

1 1 1 → 3-negative zeros on the left side

B. The potential zero's for the polynomial are: $\frac{a_0}{a_n} = \frac{-18}{3} = \frac{\pm 18}{\pm 3}$

$$= \frac{\pm 1, \pm 2, \pm 3, \pm 6, \pm 9, \pm 18}{\pm 1, \pm 3}$$

Then the potential zero's are: ±1/3, ±2/3, ±1, ±2, ±3, ±6, ±9, ±18

C. First we test the zeros: $p(1) \neq 0$, $p(-1) \neq 0$, but → $p(-2) = 0$, then we start dividing by (–2)

$$\begin{array}{r|rrrrr} -2 & 3 & 5 & 25 & 45 & -18 \\ & & -6 & 2 & -54 & 18 \\ \hline & 3 & -1 & 27 & -9 & 0 \end{array}$$

Solving → $3x^3 - x^2 + 27x - 9 = 0$ Solve by grouping gives:

$x^2 (3x - 1) + 9(3x - 1) = 0$

$(3x - 1)(x^2 + 9) = 0$

$3x - 1 = 0$ → $x = 1/3$

Or $x^2 + 9 = 0$ → $x = \pm 3i$

Then the zero's of p(x) are: {-3i, 3i, -2, 1/3} and the polynomial p(x) in factor form is:

$P(x) = 3(x+3i)(x-3i)(x+2)(x-1/3)$

2.4 Polynomial Inequalities

Steps of Solving polynomial Inequalities	Suppose that the polynomial is in one of the following forms: $F(x) < 0$; $F(x) > 0$; $F(x) \leq 0$; $F(x) \geq 0$ To solve first locate the zero's of F(x), use the zero's to divide the real numbers line into intervals.

Example -13	Solve the polynomial inequality: $x^3 \leq 3x^2$

Solution: $x^3 \leq 3x^2$

As mentioned in the steps above the inequality has to be as one of the forms shown above Or with zero on the right side: Then subtract $3x^2$ from both sides:

$x^3 - 3x^2 \leq 0$ factor this step

$x^2 (x - 3) \leq 0$

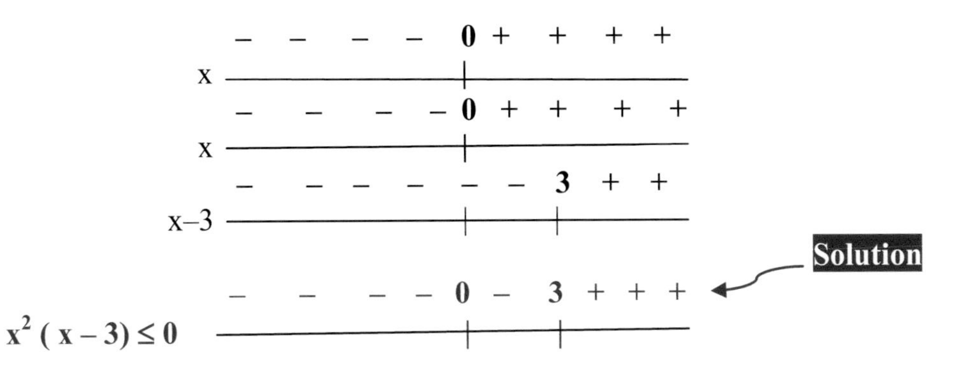

Since the solution must be negative, the solution falls between the regions: $(-\infty, 3]$

Practice -13	Solve the polynomial inequality: $x^3 > x^2$

Example -14	Solve the polynomial inequality: $(x–2)(x–3) \leq 0$

Solution: $(x–2)(x–3) \leq 0$

x–2: – – – – – 2 + + + +
(0 marked on number line)

x–3: – – – – – – 3 + + +
(0 marked on number line)

$(x–2)(x-3) \leq 0$: + + + + + 2 – 3 + + + ← **Solution**

Since the solution must be negative and closed interval, then the solution falls between
The regions: [2, 3]

Practice - 14	Solve the polynomial inequality: $3x^2 < -9x$

2.5 Application of Polynomials

Polynomials can be used to solve a variety of problems in real live, we present
Some basic ones in this section.

Example -15	Write the polynomial that represents the volume of an open box with length = 16 – x, width = 8–x, and height = x

Solution: Since volume of the box is = length x width x height
Then the volume = $(16-x)(8-x)x$

$$\text{Volume} = x^3 - 24x^2 + 128x$$

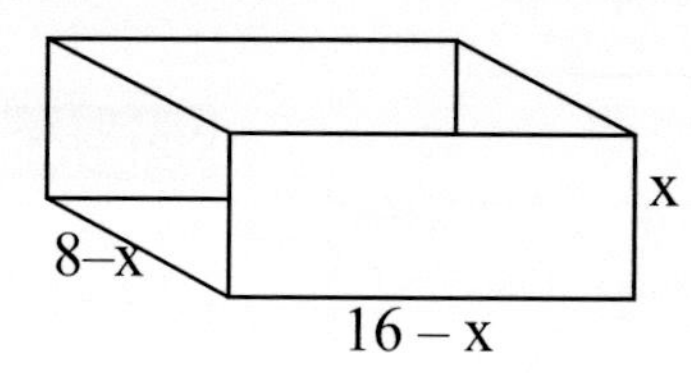

Practice -15	Write the polynomial that represents the volume of an open box with length = $10 - x$, width = $4 - x$, and height = x

Example -16	Write the polynomial that represents the area of the shaded region in the given graph.

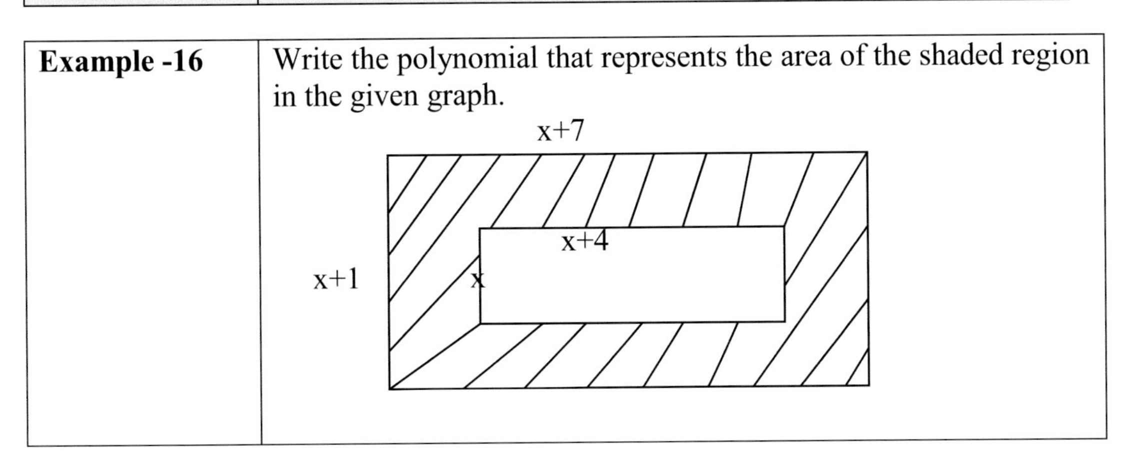

Solution: The shaded area is = Large area – small area

$$= (x+1)(x+7) - x(x+4)$$
$$= x^2 + 8x + 7 - x^2 - 4x$$
$$= 4x + 7$$

Practice -16	Express the area of the plane as a polynomial in the standard form:

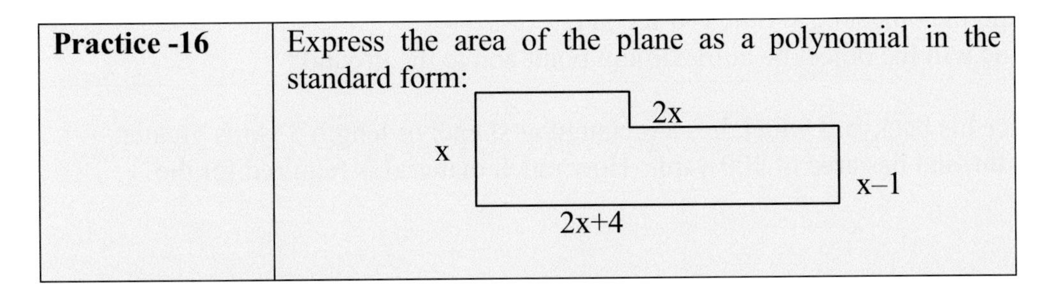

Chapter - 2 Exercise

Use synthetic division to find the quotient and remainder:

1. $3x^3 + 2x^2 - x + 3$ divide by $x-3$
2. $2x^4 - x^3 + 2x - 1$ divide by $x - \frac{1}{2}$
3. $x^5 + 8x^4 + 17x^3 + 13x^2 + 12x - 12$ is divided by $(x+5)$

Use synthetic division and the remainder theorem to find the indicated function value:

4. $f(x) = 4x^3 + 5x^2 - 6x - 4$; $f(-2)$
5. $f(x) = 2x^5 - 3x^4 + x^3 - x^2 + 2x - 1$; $f(2)$

6. Use the Rational Zero's Theorem to find all the real zeros of the polynomial function:
   $F(x) = 2x^3 - 11x^2 + 2x + 25$

Solve the polynomial inequalities, and write the solution in interval notations:

7. $(x-3)(x+1) > 0$

8. $x^2 - 6x + 8 \leq 0$
9. $x^3 \geq 6x^2$

10. A ball is projected in the air with initial velocity of 20ft/s from the ground, its height is Modeled as: $y(t) = -16t^2 + 80t + 120$.
    a. Where would the object be at time t=5 seconds?
    b. At what time will the object be at 40 ft above the ground?

11. Mark wants to fence his backyard which has a rectangular shape of length 3 yards greater than the width with area of 180 yard^2. How much material does he need to fence it?

Chapter - 2 Test

1. Use synthetic division and the remainder theorem to find the indicated function value: $f(x) = 3x^3 - 7x^2 - 2x + 5$; $f(-3)$
2. Solve the polynomial inequalities and graph the solution: $(x-3)(x+2)(x-5) \geq 0$

3. A ball is projected in the air with initial velocity of 40ft/s from the ground, its height is Modeled as: $y(t) = -16t^2 + 90t + 180$.
    a. Where would the object be at time t=6 seconds?
    b. At what time will the object be at maximum point above the ground?

4. Mark wants to fence his backyard which has a rectangular shape; of length 3 yards greater Than twice the width, and has area of 200 yard^2. How much material is required for the Fence?

3. Rational Functions

Chapter – 3
Rational Functions

Objectives: 3.1 Rational Functions and Graphs
3.2 Domain of Rational Functions
3.3 Asymptotes of Rational Functions
3.4 Rational Inequality
3.5 Applications

3.1 Rational Functions and Graphs

A rational function R(x) is a fraction form of both of its numerator and denominator are polynomials p(x), and q(x), and can be written in standard form as:

$$R(x) = \frac{P(x)}{q(x)} = \frac{a_n x^n + a_{n-1} x^{n-1} + \ldots + a_1 x^1 + a_0}{b_m x^m + b_{m-1} x^{m-1} + \ldots + b_1 x^1 + b_0}, \text{ where } q(x) \neq 0$$

3.2 Domain of Rational Functions

Example -1	Find the domain of the rational function: $F(x) = \frac{x-1}{2x^2 + 5x - 3}$

Solution: the domain of the rational functions is determined by its nonzero denominator only.

$$F(x) = \frac{x-1}{2x^2 + 5x - 3}$$

$2x^2 + 5x - 3$ $\neq 0$ gives the domain of F(x)

$$2x^2 + 5x - 3 \neq 0$$

Factor this polynomial: $(2x-1)(x+3) \neq 0$ → then $2x - 1 \neq 0$ or $x+3 \neq 0$
Then the domain of F(x) is $\{x/\ x \neq -3,\ x \neq 1/2\}$ or $\{(-\infty, -3) \cup (-3, 1/2) \cup (1/2, \infty)\}$

This can also be expressed verbally as:
The domain of F(x) is all the real numbers except –3, and 1/2.

Practice -1	Find the domain of the rational function: $F(x) = \dfrac{x+3}{x^2 - 2x - 15}$

Example-2	**Using Transformation to Graph Rational Functions**

Graph the rational Function:

$$R(x) = \frac{1}{(x+2)^2} + 1$$

To graph using Transformation method, we have to analyze the rational in the following steps:

1. Figure out what the original function was: in this function it was $p_1(x) = 1/x^2$, where $x \neq 0$, this has the following graph:

   $p(x) = 1/x^2$

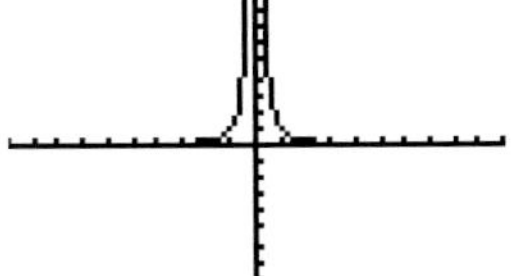

2. Shift the graph 2-units to left:

   $P(x) = 1/(x+2)^2$

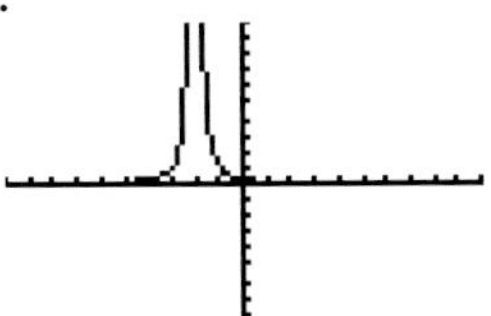

3. Shift the graph 2-units up:

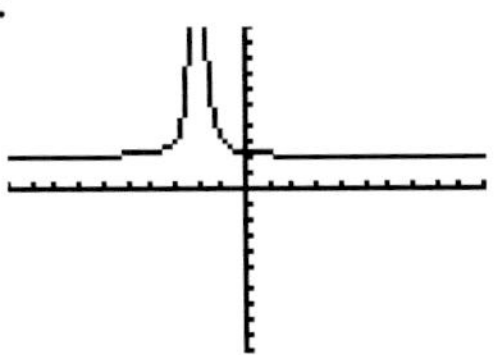

3.3 Asymptotes of Rational Functions

Asymptotes are:

a. Vertical Asymptotes.

b. Horizontal Asymptotes

c. Oblige Asymptotes.

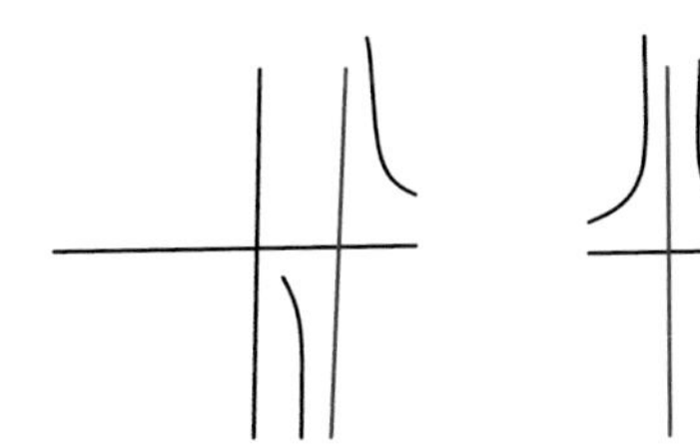
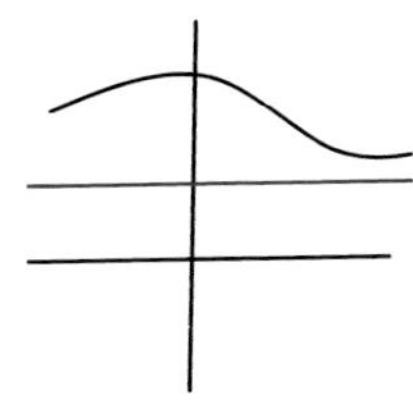
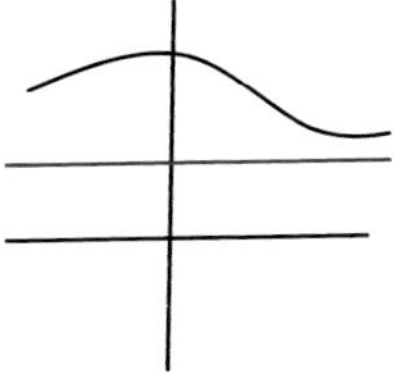
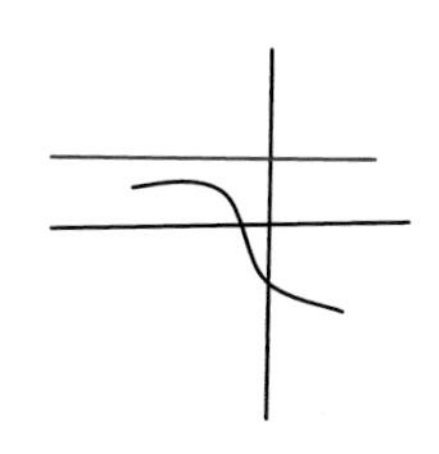

Vertical Asymptotes

Horizontal Asymptotes

Steps of Finding the asymptotes for a Rational Function

To find the asymptotes of the following Rational Function:

$$R(x) = \frac{P(x)}{q(x)} = \frac{a_n x^n + a_{n-1} x^{n-1} + \ldots + a_1 x^1 + a_0}{b_m x^m + b_{m-1} x^{m-1} + \ldots + b_1 x^1 + b_0}, \text{ where } q(x) \neq 0$$

1. Vertical asymptotes:

First we simplify the rational to its lowest term, then let the denominator x-r → o and solve for x=r is the vertical asymptote.

2. Horizontal Asymptote:

a. If n<m → the rational R(x) is a proper function, and the Horizontal Asymptote HA is y=0 or the x-axis.

b. If n=m → then the horizontal asymptote is → $y = a_n / b_m$

c. If n > m by 1 → then Long division will give oblique asymptote → y= ax+b.

d. If n> m by 2 then the long division will result in a parabola. Or there will be no asymptote

Finding Vertical Asymptotes

Exaple-2: Find the vertical asymptotes

For the rational function: $R(x) = \frac{x+7}{x^2+16}$

Solution: Here the domain is all the real numbers or $(-\infty, \infty)$ → There is no vertical asymptote

Exaple-3: Find the vertical asymptotes

For the rational function: $R(x) = \frac{x+7}{x^2-16}$

Solution: Since R(x) is in its lowest term and the domain is all the real numbers except ± 4 Or D :{ $x \neq \pm 4$} then R(x) has two vertical asymptotes at $x = -4$, and $x=4$

Exaple-4: Find the vertical asymptotes

For the rational function: $R(x) = \frac{-x^2+16}{x^2+5x+4}$

Solution: First we have to factor and simplify R(x):

$$R(x) = \frac{-x^2+16}{x^2+5x+4} = \frac{-(x-4)\cancel{(x+4)}}{(x+1)\cancel{(x+4)}} = \frac{4-x}{x+1}$$

The domain of R(x) is D:{ $x \neq -4, x \neq -1$}, but the Vertical Asymptote is $x = -1$

Finding Horizontal Asymptotes

Exaple-5: Find the horizontal asymptotes

For the rational function: $R(x) = \dfrac{-3x^2}{x^2 + 3x - 18}$

Solution: Since n = m → the horizontal asymptote is: $y = \dfrac{a_0}{b_0} = \dfrac{-3}{1} = -3$

Exaple-6: Find the oblique asymptotes

For the rational function: $R(x) = \dfrac{2x^3 + 11x^2 + 5x - 1}{x^2 + 6x + 5}$

Solution: Since n > m by 1 → then we expect to have an oblique asymptote. Long Division will give the asymptote:

$$\begin{array}{r l}
 & 2x - 1 \\
x^2+6x+5 \,\big) & 2x^3 + 11x^2 + 5x - 1 \\
 & +2x^3 + 12x^2 + 10x \quad \leftarrow \text{Change the sign of this raw} \\
\hline
 & -x^2 - 5x - 1 \\
 & -x^2 - 6x - 5 \quad \leftarrow \text{Change the sign of this raw} \\
\hline
 & x + 4 \quad \leftarrow \text{Remainder}
\end{array}$$

Dividend

$$\text{Then} \quad \frac{2x^3 + 11x^2 + 5x - 1}{x^2 + 6x + 5} = (2x - 1) + \frac{x + 4}{x^2 + 6x + 5}$$

Remainder: $x + 4$; Quotient: $(2x - 1)$; Divisor: $x^2 + 6x + 5$

The last fraction tends to zero as x increases.
Then the oblique asymptote is: $y = 2x - 1$

3.4 Rational Inequalities

This is similar what was done in polynomial inequalities

Example -7	Solve the Rational Inequality $\dfrac{x+2}{x-1} < 0$

Solution:

x+ 3: - - -3 + + + + + + +

x–1: - - - - - - 1 + + + +

$\dfrac{x+3}{x-1} < 0$: + -3 – 1 + ← **Solution**

Then the solution set is {(–3, 1)}

Practice -7	Solve the Rational Inequality $\dfrac{3x+1}{x-2} \geq 0$

Example -8	Solve the Rational Inequality $\dfrac{x+2}{x-5} > \dfrac{3}{x-5}$

Solution:

$$\frac{x+2}{x-5} - \frac{3}{x-5} > 0$$

$\dfrac{x-1}{x-5} > 0$ solving in the same manner, we find that the positive solution exists in the region in the region → {(–∝, 1) ∪ (5, ∝)}

Practice -8	Solve the Rational Inequality $\dfrac{4x+2}{x+4} \geq 1$

3.5 Applications

Example -9	Find the area of a rectangle with length= $\frac{2x}{(x+5)}$, And width = $\frac{x}{(x+3)}$.

Solution: Area of rectangle = Length • Width

$$= \frac{2x}{(x+3)} \bullet \frac{x}{(x+5)} = \frac{2x^2}{(x+3)(x+5)}$$

Practice -9	Find the area of a rectangle with length= $\frac{3x}{(x+1)}$, And width = $\frac{x}{(x+2)}$.

Chapter - 3 Exercise

Solve the following rational inequalities and graph the solution:

1. $\frac{x-4}{x+2} \leq 0$

2. $\frac{x^2 - 3x + 2}{x^2 - 2x - 3} \geq 0$

3. $\frac{3}{x+2} > \frac{2}{x-2}$

5 . $\frac{(x+12)(x-2)}{(x-1)} \geq 0$

6 . $\frac{8x-3}{x+4} \leq 7$

A company manufactures bicycles for kids. The average cost function of their production Was modeled as:

$$C(x) = \frac{6000 + 300x}{x}$$

Where, C(x) is the average cost of production in dollars, and x is the number of bicycles Produced Per month:

7. What is the average cost of production of 100 bicycles per month?
8. How many bicycles should the company produce so that the cost does not exceed $350?

Chapter - 3 Test

1. Solve the rational inequality:

$$\frac{1}{4(x-1)} \geq \frac{4}{(x+3)}$$

2. Solve the rational inequality

$$\frac{3x+6}{x^2-6x+5} \geq 0$$

4. Exponential and Logarithmic Functions

Chapter – 4
Exponential and Logarithmic Functions

Objectives: 4.1. Introduction
4.2. One-to-one Functions
4.3. Exponential Functions
4.4. Logarithmic Functions
4.5. Properties of Logarithmic Functions
4.6. Logarithmic and Exponential Equations
4.7. Applications

4.1 Introduction

Logarithmic Functions	The logarithm function of x to base b is written in the form: $\log_b x$ $y = \log_b x \rightarrow \text{IFF } x = b^y$
Exponential Functions	An exponential function is a function that can be expressed in the form, $f(x) = b^x, b > 0, b \neq 0$ b is the base of the exponential function, and base of the logarithmic function too. Logarithmic functions and exponential functions are inverse of each other: $y = \log_b x \longrightarrow x = b^y$

The graph below shows how the logarithmic functions and exponential functions are Inverse of each other:

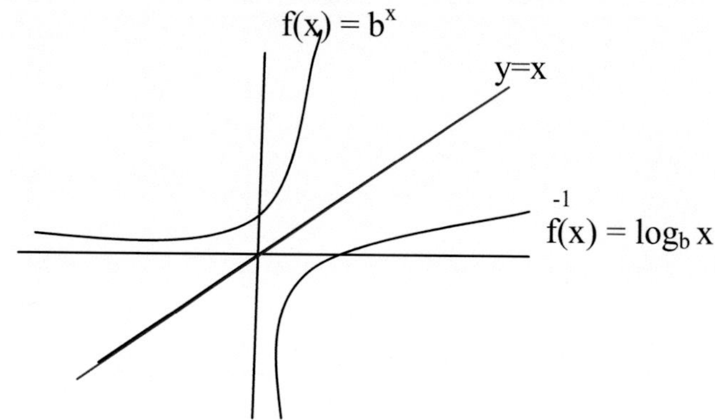

Example - 1	Graphing Exponential and Logarithmic Functions

For the exponential function: $f(x) = 2^x$:
A. Find the domain, and the end behavior
B. Find the inverse function (logarithmic)
C. Graph them both in one graph.

Solution:

A. Domain of the exponential function is $(-\infty,\infty)$.

The end behavior: As $x \rightarrow \infty$, $f(x) = 2^x$ grows faster, and rise rapidly.

As $x \rightarrow -\infty$ the value of $f(x) = 2^x$ gets closer to zero.

Then the x-axis (y=0) is the horizontal asymptote for $f(x) = 2^x$

B. To find the inverse: first we rewrite f(x) as y:

$y=2^x$ then replace $y \leftrightarrow x$

$x = 2^y$ then solve for y

$\log x = \log 2^y$

$$\log x = y \log 2 \rightarrow y = \frac{\log x}{\log 2} = \log_2 x$$

C.

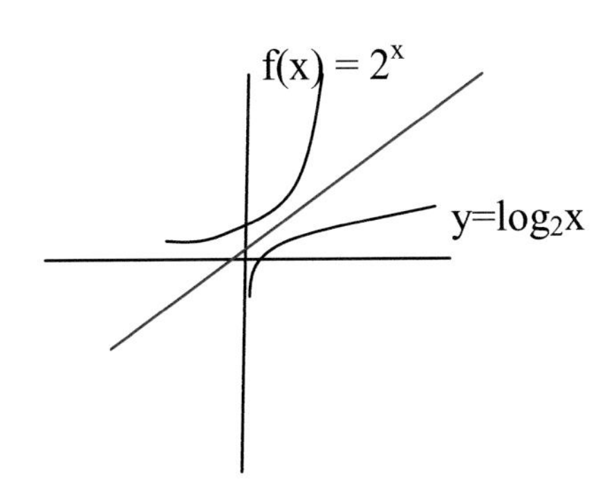

Example - 2	Graphing Exponential Function

Graph the following exponential functions:
A. $y=2^x$
B. $y= 3^x$
C. $y=(1/2)^x$
D. $y= (1/3)^x$

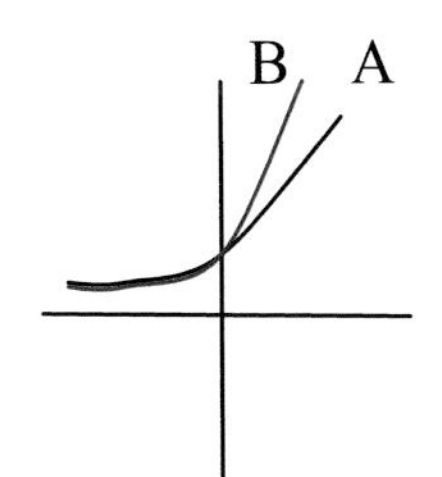

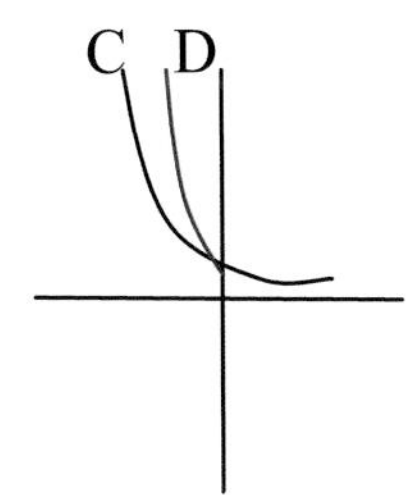

4.2 One-To-One Functions

Definition	A function is one-to-one function if any two different inputs in the domain corresponds to two different outputs in the range, that is if x_1, x_2 are two different inputs of a function f, then $f(x_1) \neq f(x_2)$
Definition of 1-1 function	To test if a function is one-to-one function a horizontal line test is used. If any horizontal line intersects the graph of a function in at most one point, then the function is 1-1 function.
Inverse Functions	Inverse functions are 1-1 functions, the inverse of f(x) is $f^{-1}(x)$ Domain of f(x) = Range of $f^{-1}(x)$ Range of f(x) = Domain of $f^{-1}(x)$
How to find the inverse	To Find the inverse replace x with y, and to find the inverse of a function algebraically, exchange y with x, then solve for y that is the inverse.

Example -3	Determine if the functions are one-to-one functions: a. $f(x) = x^2$ b. $f(x) = x^3$

Solution: **a.** the graph of the function is a parabola opened up
if we let $x_1 = -1$, and $x_2 = 1$ Then substituting in gives:
$f(x_1) = f(-1) = (-1)^2 = 1$
$f(x_2) = f(1) = (1)^2 = 1$
Since f(x) = f(x) → then the function f(x) = x is not a 1–1 function as shown.

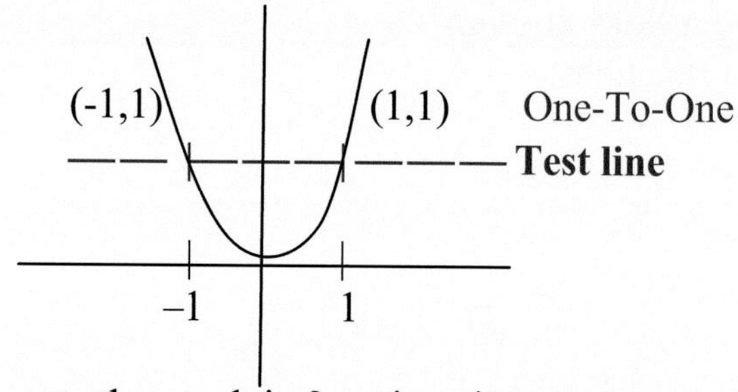

The test line intersects the graph in 2=points, it means not 1–1
b. Choosing the same two points on the domain, and find their f(x):
$f(x_1) = f(-1) = (-1)^3 = -1$
$f(x_2) = f(1) = (1)^{23} = 1$
Since $f(x_1) \neq f(x_2)$ → then the function $f(x) = x^3$ is a 1–1 function.

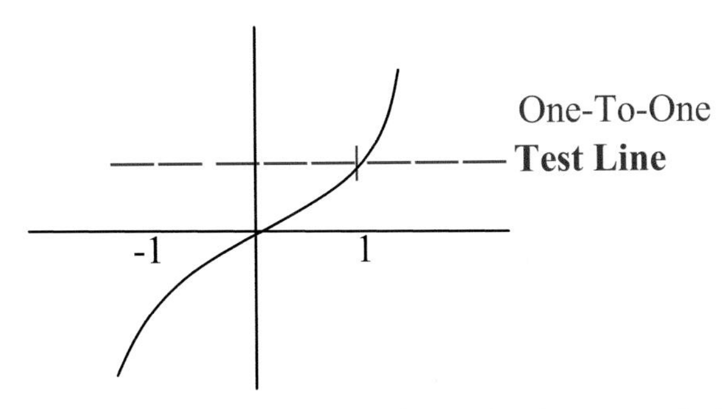

The test line intersects the graph in one point only, it means 1–1

Practice -3	Determine if the graph of the following functions are 1–1: 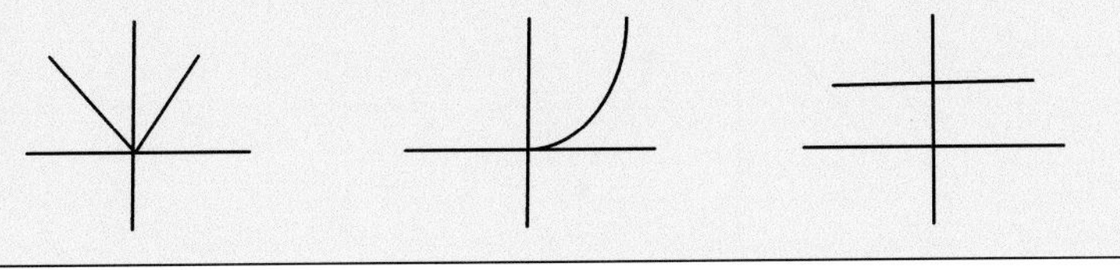

Example -4	Find the in inverse of the following set of ordered pairs: $f(x) \rightarrow \{ (-2,-8), (-1,-10), (1,1), (2,7), (3,15) \}$

Solution: we exchange the x-values with y values to get the inverse function.

$f(x) \rightarrow \{ (-2,-8), (-1,-10), (1,1), (2,7), (3,15)\}$
$f^{-1}(x) \rightarrow \{ (-8,-2), (-10,-1), (1,1), (7,2) ,(15,3) \}$

Practice -4	Find the inverse of the following sets: $f(x) = \{ (-10,9), (-7,8), (-4, -2), (2,5) ,7,10) \}$

Obtaining a graph of $f^{-1}(x)$ from graph of f(x):

Example -5	Find the in inverse of the given graph: 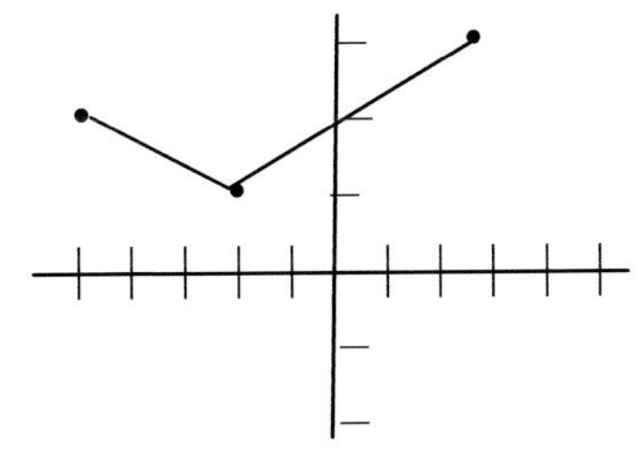

Solution: To find the inverse, we will find the inverse of the points then graph it.

f(x) points	$f^{-1}(x)$ points
–5,2	2,–5
–2, 1	1,–2
3,3	3,3

Then the graph of f(x) and $f^{-1}(x)$ is as follows:

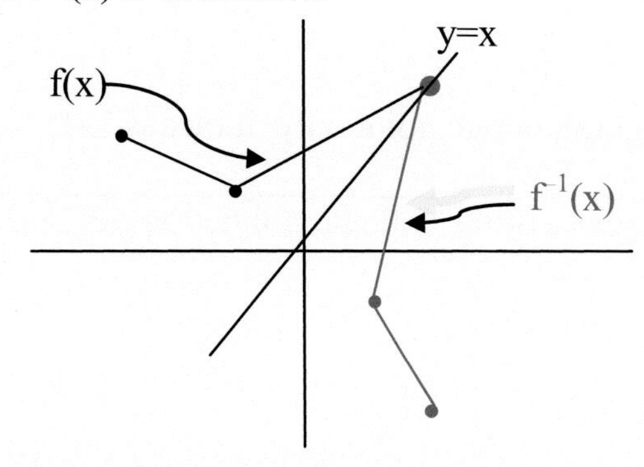

Finding the Inverse of f(x) Algebraically

Example -6	Find the inverse of function algebraically: f(x) = 3x –1

Solution: Let f(x) = y

y = 3x –1, then replace y with x

x = 3y –1, now solve for y:

x+1= 3y → then y = 1/3(x+1) = $f^{-1}(x)$

The graph of both f(x) and $f^{-1}(x)$ is as given:

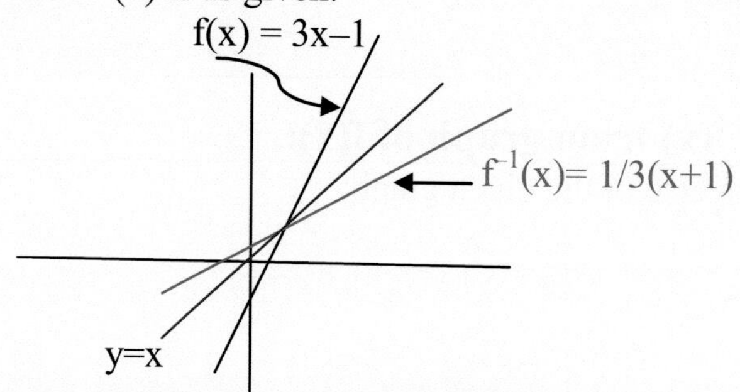

Practice -6	Find the inverse of function algebraically: $f(x) = \frac{2x+3}{x+1}$

Example -7	Find the inverse of function algebraically: $f(x) = x^3 - 1$

Solution: Replace f(x) with y

$y = x^3 - 1$, then replace y with x

$x = y^3 - 1$

$y^3 = x+1$ Take radical 3 on both sides, we get

$y = \sqrt[3]{x+1} = f^{-1}(x)$

Practice -7	Find the inverse of the function algebraically: $f(x) = x^3 - 8$

Example -8	Find the inverse of the function algebraically: $f(x) = (x-2)^2$

Solution: Replace f(x) with y

$y = (x-2)^2$, then replace y with x

$x = (y-2)^2$, now take the square root of both sides

$\sqrt{x} = y-2$ Solve for y → $y = \sqrt{x} + 2 = f^{-1}(x)$

Practice - 8	Find the inverse of the function algebraically: $f(x) = \frac{x}{4} + 2$

Convolution	If two functions f(x), and g(x) are inverse of each other then: $f(g(x)) = g(f(x)) = x$

Example -9	Show that the two functions $f(x) = (x-2)^2$, and $g(x) = \sqrt{x} + 2$ are inverse of each other.

Solution: If f(x) is inverse of g(x) then: $f(g(x)) = g(f(x)) = x$

L.S → $f(g(x)) = f(\sqrt{x} + 2) = (\sqrt{x} + 2 - 2)^2$

$= (\sqrt{x})^2 = x$

R.S → $g(f(x)) = g((x-2)^2) = \sqrt{(x-2)^2} + 2 = X - 2 + 2 = x$

Practice-9	Show that the two functions $f(x) = 4x-8$, and $g(x) = x/4 + 2$ are inverse of each other.

4.3 Exponential Functions and graphs

Definition	Exponential functions are of the form: $f(x) = a^x$ a is a positive real number, x is a variable if $a > 1$ → f(x) is an increasing function if $0 < a < 1$ → f(x) is decreasing function
Properties of Exponential functions	• Domain is all the real numbers, range is $y > 0$. • x-axis is the horizontal asymptote • No x-intercepts, and y-intercept is 1 • graph is smooth and continuous

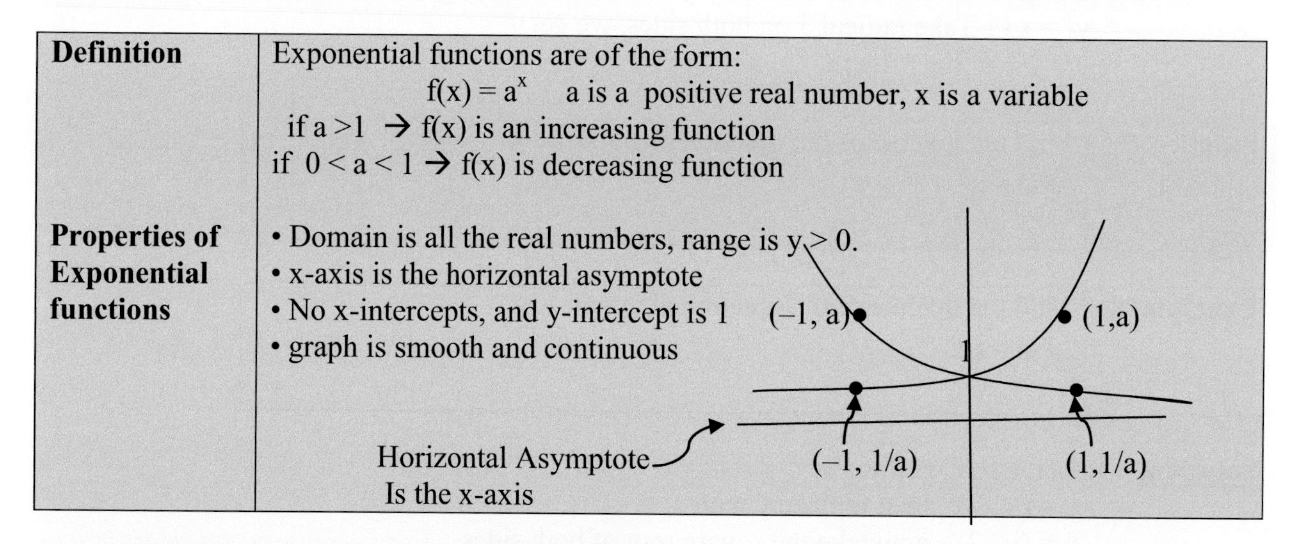

4.4 Logarithmic Functions and Graphs

Definition	$y = \log_b x$ IFF → $x = a^y$
Properties of Logarithmic functions	• Range is all the real numbers, Domain $x > 0$ • y-axis is the vertical asymptote • No y-intercepts, and x-intercept is 1 • graph is smooth and continuous
Logarithmic functions and Exponential functions	

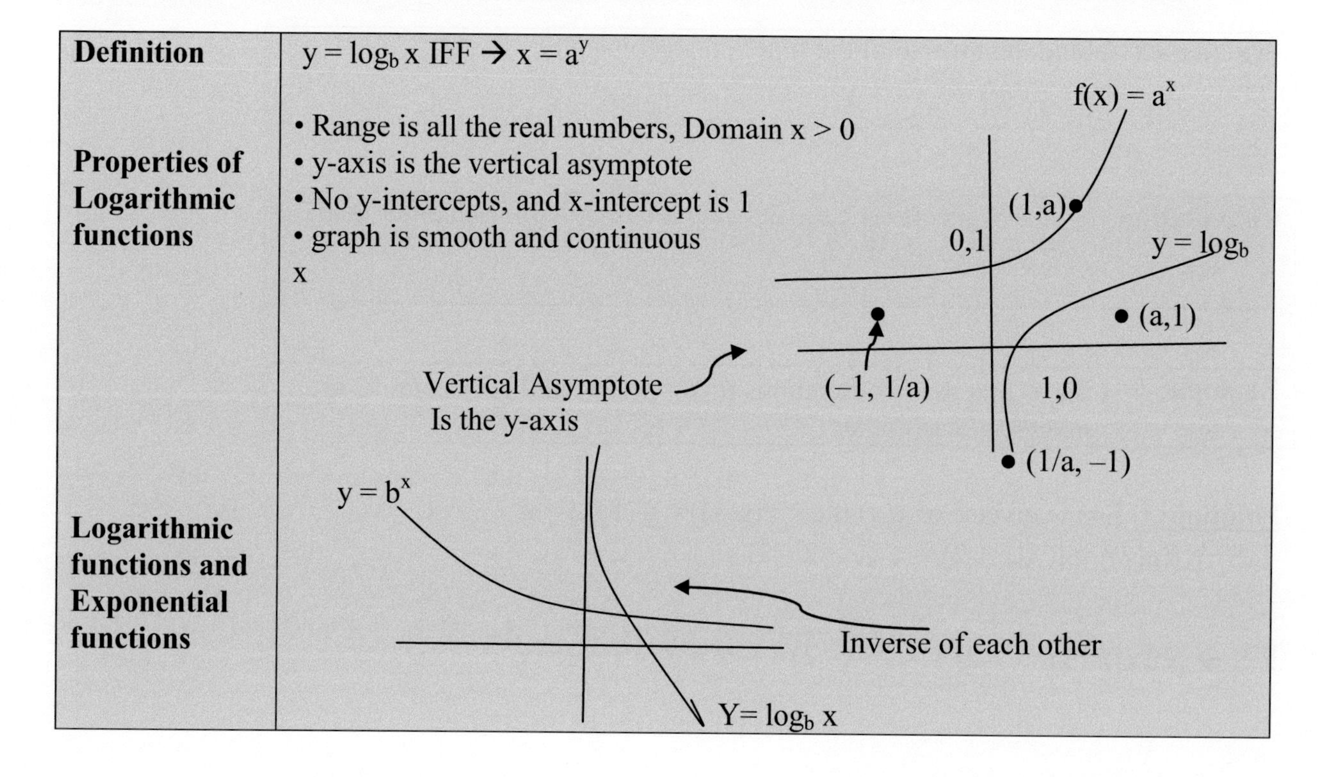

Example -10	Graph the logarithmic function and check the properties: $f(x) = \ln(x-2)$

Solution: The graph shows the function shifted to the right by 2-units

The domain of the function is x>2

Practice-10	Graph the following exponential functions and check the properties: a. $f(x) = 2^x$ b. $f(x) = 3^x$ c. $f(x) = 4^x$ d. $f(x) = (1/2)^x$ e. $f(x) = (1/4)^x$ f. $f(x) = (1/5)^x$

4.5 Properties of Logarithmic Functions

$\log_b b = 1$

$a^{\log_a x} = x$

$\log(xy) = \log x + \log y$

$\log_b \frac{x}{y} = \log_b x - \log_b y$

$\log_b x^n = n \log_b x$

If $x = y$ → then $\log_b x = \log_b y$

If $a^x = a^y$ → then $x = y$

Change of Base Rule: $\log_b x = \dfrac{\log x}{\log b} = \dfrac{\ln x}{\ln b}$

4.6 Solving Logarithmic and Exponential Functions

Example-11	Solve the exponential equation: a. $5^x = 5^2$ b. $(1/3)^x = 27$

Solution: Using the above rule gives a. → x=2

b. $(1/3)^x = 3^3$

$(1/3)^x = (1/3)^{-3}$ → $x = -3$

Practice - 11	Solve the exponential equation: a. $7^{2x} = 7^2$ b. $(1/4)^x = 32$

Example-12	Solve the exponential equation: a. $4^{5x-3} = 16^{x+2}$ b. $e^2 = e^{3x+6}$

Solution: a. $4^{5x-3} = 16^{x+2}$

$4^{5x-3} = (4)^{2(x+2)}$

Then, $5x-3 = 2x+4 \rightarrow 3x = 7 \rightarrow x = 7/3$

b. $2 = 3x + 6 \rightarrow 3x = -4 \rightarrow x = -4/3$

Practice - 12	Solve the exponential equation: a. $5^{1/2x} = 5^2$ b. $e^x = e^{7x+2}$

Example-13	Solve the exponential equation: $3^x \bullet 9^{x^2} = 81^x$

Solution: $3^x \, . \, 9^{x^2} = 81^x$

$3^x \, . \, 3^{2x^2} = 3^{4x}$

$3^{x+2x^2} = 3^{4x}$

$x + 2x^2 = 4x$

$2x^2 - 3x = 0$

$x(2x-3) = 0 \rightarrow x=0$, or $x=3/2$ the solution set is $\{0, 3/2\}$

Practice-13	Solve the exponential equation: $\frac{3^{2x}}{3^{x^2}} = 27$

Solving Logarithmic Equations

Example-14	Solve the Logarithmic equation: $5 \ln(2x) = 30$

Solution: $5 \ln(2x) = 30$

$\ln(2x) = 30/5 = 6 \rightarrow$ but $\ln = \log_e$

Then $\rightarrow \log_e(2x) = 6 \rightarrow 2x = e^6 \rightarrow x = 1/2\, e^6$

Check: substituting back $\rightarrow$ L.S $= 5 \ln(2 \,.\, 1/2\, e^6) = 5 \ln(e^6)$

$= 5.\log_e e^6 = 30 =$ R.S

Practice-14	Solve the following Logarithmic equations: a. $\log(5+x) - \log_6(x-3) = \log 5$ b. $\log_3(x+1) + \log_3(x-5) = 3$

4.7 Applications

Logarithmic and exponential functions are used in modeling different type of problems in life such as:

a. probability problems(exponential probability)
b. Business problems.(Population growth, compound Interest)
c. Electronic problems (RL Circuit)
d. Chemistry problems (Bacteria)

Exponential Decay and Growth

The function: $p(t) = p_0 e^{-kt}$, $k>0$ is used for decay, and
The function: $p(t) = p_0 e^{kt}$, $k>0$ is used for growth.
where:
p_0 represents the amount of substance at time $t=0$.
$P(t) =$ the amount of the substance after time t.
$k =$ positive constant called the decay rate.
The first function is a decreasing function, while the second one is the increasing function as shown in the graph:

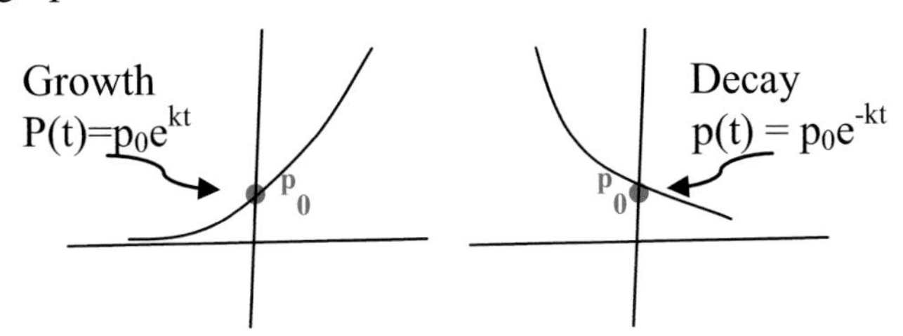

Example - 15	Bacterial Growth

The bacterial growth in medicine is modeled by the exponential equation:

$$N = N_0\, e^{1.386\,t} \quad ......(1)$$

Where, N=the number of bacteria present at time t.
N_0 = the number of bacteria present at time t=0.
If we start with one-bacterium, how many will be present after:
A. 3 hours.
B. 10 hours.

Solution: Using the above formula (1) gives:

A. N(after 3 hours) = 1. $e^{1.386(3)} \approx 64$

B. N(after 10 hours) = 1. $e^{1.386(10)} = 1045493.94$

Compound Interest

The compound interest formula is:

$$A = p\left(1 + \frac{r}{m}\right)^{mt} \quad .... (2)$$

Where, A = future amount
P= principal
r= rate per year in percent
m= the compounding period
t = time in years

m	Payment period
1	Annually
2	Semi-annual
4	Quarterly
12	Monthly
52	Weekly
360	daily

Example - 16	Compound Interest

Find the amount that result from each investment after 3-years:
A. \$1000deposit in the bank with interest rate of 7% compounded monthly.
B. \$1000deposit in the bank with interest rate of 7% compounded quarterly
C. \$1000deposit in the bank with interest rate of 7% compounded yearly

Solution: using the above formula (2):

A. $A = 1000(1 + 7\%/12)^{12(3)} = \$1232.93$
B. $A = 1000(1 + 7\%/4)^{4(3)} = \$1231.44$
C. $A = 1000(1 + 7\%/1)^{(3)} = \$1225.04$

Example - 17	Compound Interest

What rate of interest compounded semi-annually is required to double an investment in 3-years?

Solution: Here we have: r is missing, m=2, if we let the principal = p, then a = 2p, then using formula (2):
$2p = p(1 + r/2)^{2(3)}$

$2 = (1+r/2)^{2(3)}$

$2^{1/6} = 1 + r/2 \rightarrow r = 2(2^{1/6} - 1) = \$24.49\%$

Another type of compounding is the continuous compounding with formula:

$A = Pe^{rt}$ .... (3)

Where: A = future amount
r = interest rate
t = time in years
p = principal

Example - 18	Compound Continuously

How long does it take for an investment to double in value with rate of 7% compounded continuously?
Solution: Using formula (3) with: A = 2p, and r=7%, we need to find t?

$2p = p\,e^{7\%(t)} \rightarrow 2 = e^{0.07\,t} \rightarrow \ln 2 = (0.07)\,t \ln e$

Then $\rightarrow t = \ln 2/(0.07) = 9.9$ years.

Chapter - 4 Exercise

Determine if the function is one-to-one function:

1. a. Domain Range

b. Domain Range

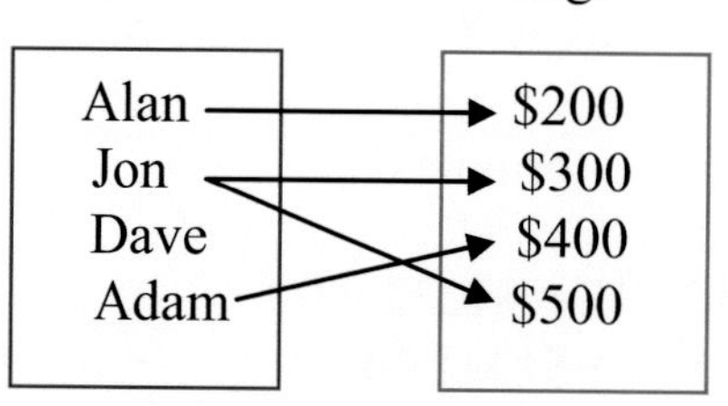

2. {(2,1), (3,3), (4,5), (2,10) }

3. using the horizontal-test-line determine if the graph is one-to-one function:

a.

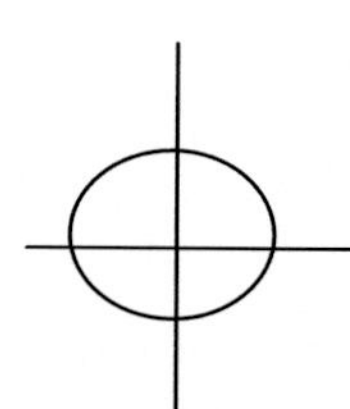

b.

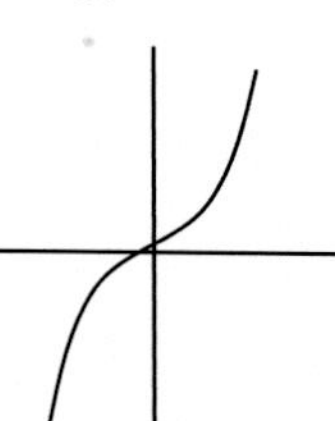

c.

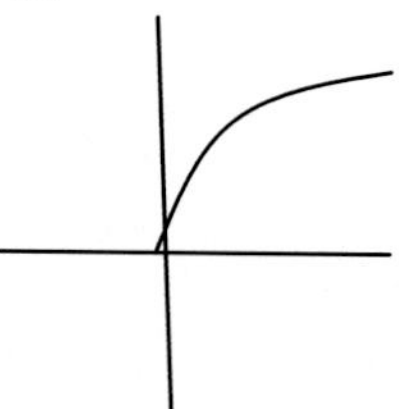

Verify that the functions f, and g are inverse of each other by showing that: $f(g(x)) = g(f(x)) = x$

4. $f(x) = \dfrac{3x + 5}{1 - 2x}$ $\quad g(x) = \dfrac{x - 5}{2x + 3}$

5. $f(x) = 5x - 1$ $\quad g(x) = 1/5(x+1)$

6. $f(x) = \sqrt{x + 8}$ $\quad g(x) = x^2 - 8$

7. A graph of function f is given; draw the inverse f^{-1}:

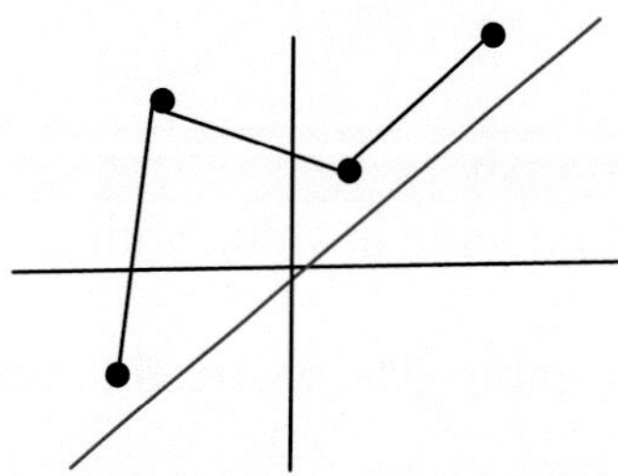

Find the inverse of the following functions algebraically:

8. $f(x) = \dfrac{2x + 1}{x - 1}$

9. $f(x) = \sqrt{x+4}$

10. $f(x) = \sqrt{x} + 4$

11. Change the exponential expression to an equivalent expression involving a logarithm:
$4^{5/2} = 32$

12. Change the logarithmic expression to an equivalent expression involving exponent:
$\text{Log}_b\, 125 = 3$

13. Find the exact value of the logarithmic expressions:
a. $\text{Log}_{1/3}\, 9$ b. $\ln e^{12}$

14. Write as the sum or difference of logarithms: $\log_2 \dfrac{17\sqrt{m}}{N}$

15. Express as a single logarithm: $3 \log_6 x + 5 \log_6 (x-6)$

16.Solve the logarithmic equation: $\log(5+x) - \log(x-3) = \log 5$

Chapter - 4 Test

1. Determine if the function is one-to-one function:
a. {(1, 2), (3, 5), (6, 7), (10, 12)}
b. {(3, 2), (7, 2), (9, 1), (3, 5)}
c. {(-2, 1), (0, 2), (3, 4), (5, 4)}

2. Use the horizontal-test-line to determine if given function is one-to-one function:

a.

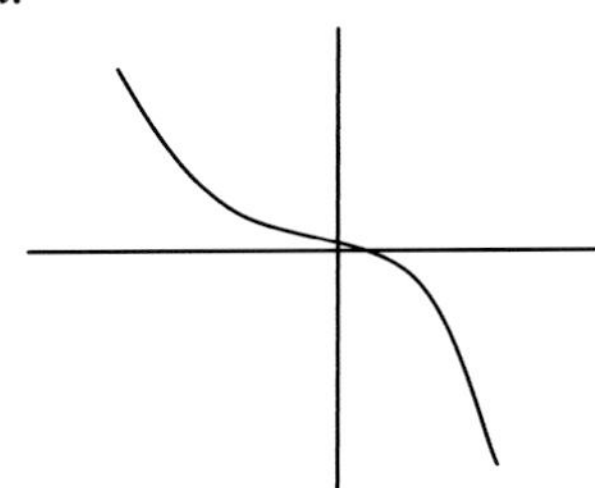

b.

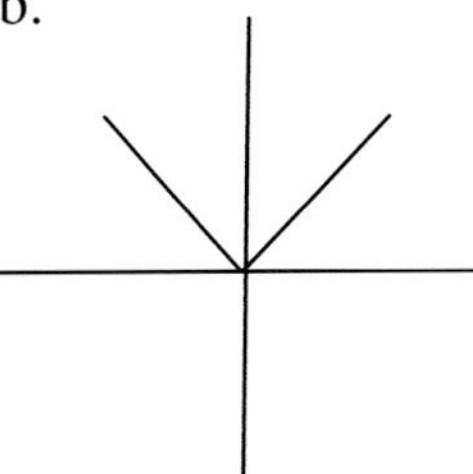

3. Are f, and g inverse of each other?
a) $f(x) = 7x + 1$; $g(x) = x+1$

b) $f(x) = \sqrt{2x + 4}$; $g(x) = 1/2(x^2 - 4)$

4. Graph the inverse function f^{-1} for the given function f:

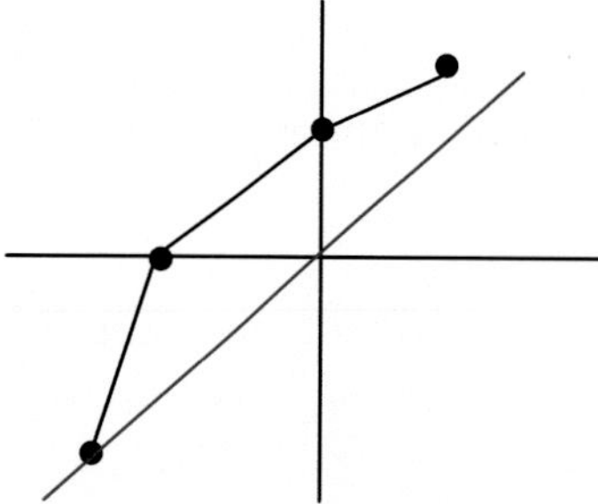

5. Find the inverse function algebraically: $f(x) = \dfrac{3x+1}{2}$

6. Write as the sum or difference of logarithms: $\log_4 \dfrac{\sqrt{x}}{64}$

7. Express as a single logarithm: $\frac{1}{2} (\log_3 (x - 5) - \log_3 x)$

8. Solve the logarithmic equation: $\log_3(x + 1) + \log_3(x-5) = 3$

9. Find the future amount for \$2000 invested at 3.5% weekly for a period of 3 years.

10. What rate of interest compounded annually is required to double an investment in 4 years?

11. How long does it take for an investment to double in value if it is invested at 7% Compounded continuously?

5. Trigonometric Functions

Chapter – 5
Trigonometric Functions

Objectives: 5.1 Angles and Their Measures
5.2 Trigonometric Functions
5.3 Solving Right Triangles
5.4 Properties of the Trigonometric Functions
5.5 Trigonometric Identities
5.6 Applications

5.1 Angles and Their Measures

The angle is formed of a fixed side (ray) called the initial ray, and a rotating side (ray) called the terminal ray. If the terminal ray is rotating in the counter clockwise direction, it forms a positive angle. But if the terminal ray is rotating in the clockwise direction then it forms a negative angle as shown in fig (1) and fig (2).

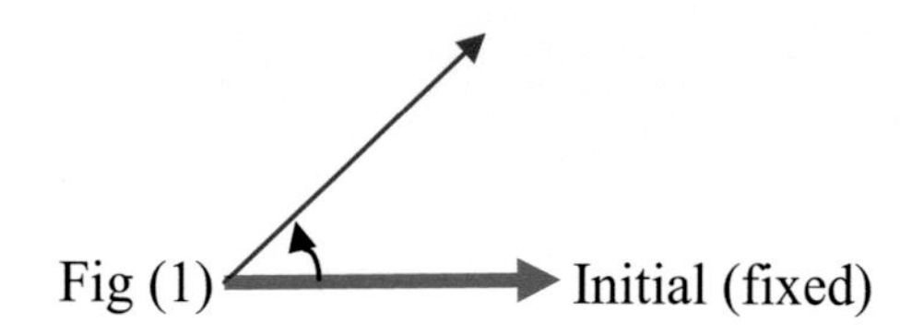

Fig (1)

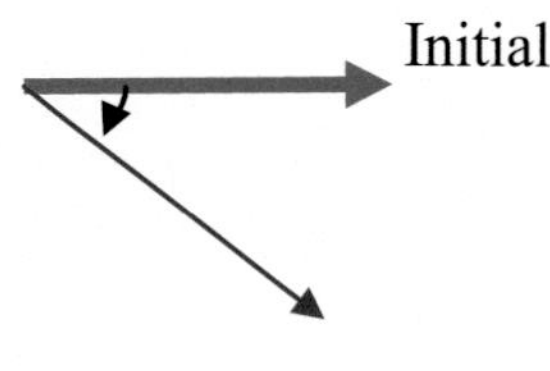

Fig (2)

The x and y axis divide the 2D space into 4-parts called quadrants labeled as Q1, Q2, Q3, and Q4 in the counter clockwise direction fig(3).

Q2 Q1
Q3 Q4

Fig(3)

Measure of Angles

Two types of measures are used for angles:

a. Degrees and sub degrees.
b. Radians.

a. The standard degrees used are: 0°, 30°, 45°, 60°, 90°, 180°, 360° for one cycle.
The sub-degrees are minutes ('), and seconds (").

b. Radians are: $\pi/6$, $\pi/4$, $\pi/3$, $\pi/2$, π, 2π.

Graphing The Standard Positive angles

The following angles represent the positive angles where the initial is fixed, and the terminal ray is moving in the counter clockwise direction: Blue arrow represents the initial ray, and the red arrow represents the terminal ray.

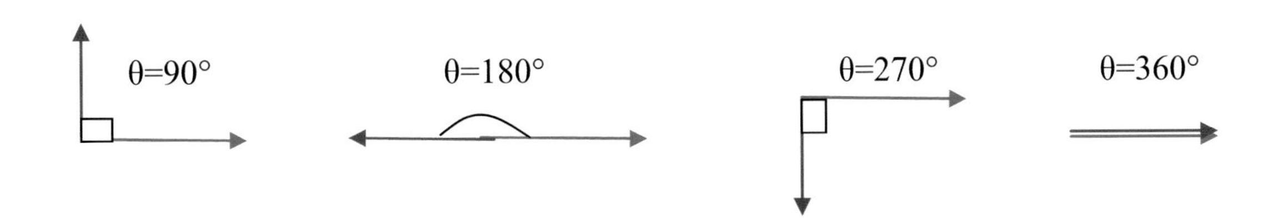

Graphing the standard negative angles:

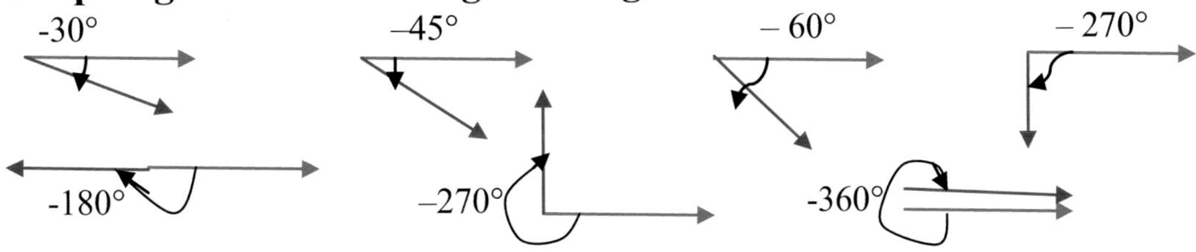

Angle	Name
90°	Right Angle
180°	Straight Angle
360°	Circular Angle

1° = 60’ 1’ = 60” 1 revolution = 360° = (360 x 60)’ = (360 x 60 x 60)”

Conversion of Angles

To convert angles from degrees to DMS we multiply by multiplicity of 60.
To convert angle from DMS we divide by multiplicity of 60.

Example-1:

a. Convert 30° 8’ 20” to decimal degrees.

b. Convert 25.325° to degrees, minutes, and seconds (DMS)

Solution: **a.** 30 8 20 = 30° + 8' + 20"

$= 30 + (8/60)° + (20/60^2)°$

$= 30° + 0.133333° + 0.005555555°$

$\approx 30.139°$

b. 25.325°= 25° + 0.325°

= 25° + (0.325 x 60)'

= 25° + 19.5'

= 25 + 19' + 0.5'

= 25° + 19' + (0.5 x 60)"

= 25° + 19 + 30"

Then DMS = 25° 19' 30"

Using TI-83: For the symbols press second → Catalog then look for the symbol

a.

30° 8' 20"
30.13888

b.

25.325°
25° 19' 30"

1 radian = π = 180° Straight Angle	2 radians = 360° = Circular angle

Arc Length

In a circle of radius r and central angle θ its length s is given as: s = r θ

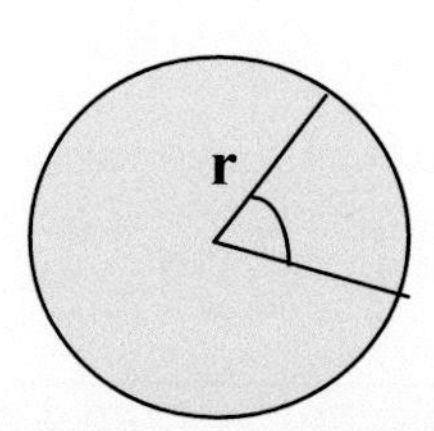

Example-2: Find the length of the arc S of the circle of radius r = 3 m, and angle θ = 0.23 radians.

Solution: S = r θ = 3 m (0.23 rad) = 0.29 m

Degrees vs Radians:

To convert from degrees to radians multiply the degree by (π/180)

To convert from radian to degrees multiply the radian by (180 / π)

Example-3: Convert the following angles from degrees to radians:

a. 50° b. 130° c. 80° d. -35°

Solution:

a.50 (π/180) = 5π/18 rad

b. 130 (π/180) = 12π/18 rad

c. 80 (π/180) = 4π/9 rad

d. -35 (π/180) = 7π/36 rad

Example-4: Convert the following angles from radians to degrees:

a.π/2 rad b. 7π/2 rad c. - 3π/4 rad d. 5π/3 rad

Solution:

a. π/2 (180/π) = 90°

b. 7π/2 (180/π) = 630°

c. –3π/4 (180/π) = –135°

d. 5π (180/π) = 300°

Table-1

Degrees	**0°**	**30°**	**45°**	**90°**	**270°**	**360°**
Radians	**0**	**π/6**	**π/4**	**π/2**	**3π/2**	**2π**

Circular Motion

An object moving on a straight line with speed (v) in time (t) covers a distance: s = v t.
Or the linear speed: v = s/t.
If the object is moving on a circular curve of radius r and angular speed ω (omega), the angular distance θ covered is given as: $\theta = \omega$ t, or the angular speed is: $\omega = \theta$/t.

Linear speed v = s/t Angular speed ω = θ / t

The Relation between Linear and Angular Speed

We have found earlier that S = r θ, and now S = v t, then:

$$r\,\theta = v\,t \rightarrow v = r\,(\theta/t) = r\,\omega$$

Then v (linear speed) = r ω (angular speed)

$$r = \frac{\text{Linear speed}}{\text{Angular speed}} = \frac{v}{\omega}$$

5.2 Trigonometric Functions

A. The Unit Circle Approach

The unit circle is a circle with radius = 1 unit.

For a unit Circle r = 1 , and S = θ

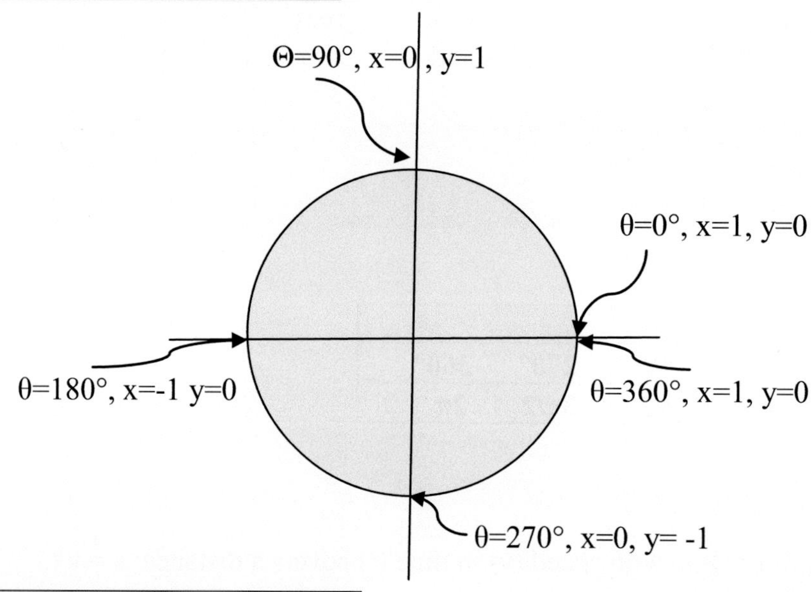

B. The Right Triangle Approach

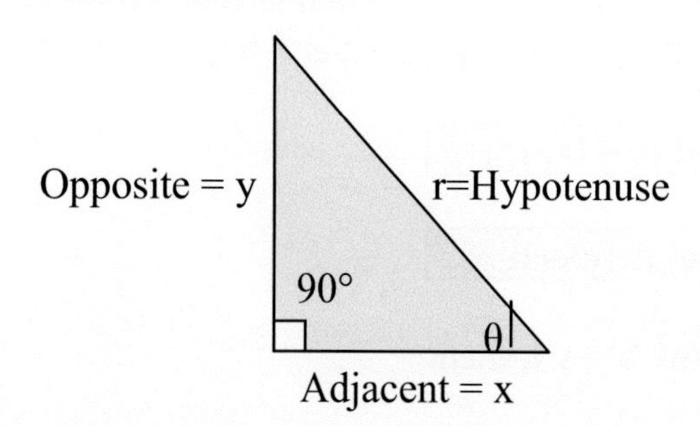

Using the right triangle above, we can write the six-Trigonometric functions as:

$$\text{Sin } \theta = \frac{\text{Opposite}}{\text{Hypotenuse}} = \frac{y}{r}$$

$$\text{Cos } \theta = \frac{\text{Adjacent}}{\text{Hypotenuse}} = \frac{x}{r}$$

$$\text{Tan } \theta = \frac{\text{Opposite}}{\text{Adjacent}} = \frac{\text{Sin } \theta}{\text{Cos } \theta} = \frac{y}{x}$$

$$\text{Cot } \theta = \frac{\text{Adjacent}}{\text{Opposite}} = \frac{\text{Cos } \theta}{\text{Sin } \theta} = \frac{1}{\text{Tan } \theta} = \frac{x}{y}$$

$$\text{Sec } \theta = \frac{\text{Hypotenuse}}{\text{Adjacent}} = \frac{1}{\text{Cos } \theta} = \frac{r}{x}$$

$$\text{Csc } \theta = \frac{\text{Hypotenuse}}{\text{Opposite}} = \frac{1}{\text{Sin } \theta} = \frac{r}{y}$$

Using the unit circle where r=1, the above six trigonometric functions will be:

Sin θ = y	**Cos θ = x**
Tan θ = y/x	**Cot θ = x/y**
Sec θ = 1/x	**Csc θ = 1/y**

Finding the exact values for the standard Trigonometric Functions for :
θ= 0°, 90°, 180°, 270°, and 360° :

1. **At θ=0° , x=1, y = 0:**

   Sin 0 = y = 0
   Cos 0 = x = 1
   Tan 0 = y/x = 0/1 = 0
   Cot 0 = x/y = 1/0 = undefined
   Sec 0 = 1/x = 1/1 = 1
   Csc 0 = 1/y = 1/0 = undefined

2. **At θ= 90°, x = 0, y=1:**

   Sin 90 = y = 1
   Cos 90 = x = 0
   Tan 90 = y/x = 1/0 = undefined
   Cot 90 = x/y = 0/1 = 0
   Sec 90 = 1/x = 1/0 = undefined
   Csc 90 = 1/y = 1/1 = 1

3. **At θ = 180°, x= –1, y= 0:**

   Sin 180 = y = 0
   Cos 180 = x = -1
   Tan 180 = y/x = 0/–1 = 0
   Cot 180 = x/y = – 1/0 = undefined
   Sec 180 = 1/x = 1/–1 = –1
   Csc 180 = 1/y = 1/0 = undefined

4. **At θ = 270°, x = 0, y= -1:**

   Sin 270 = y = -1
   Cos 270 = x = 0
   Tan 270 = y/x = -1/0 = undefined
   Cot 270 = x/y = 0/-1 = 0
   Sec 270 = 1/x = 1/0 = undefined
   Csc 270 = 1/y = 1/-1 = -1

5. **At θ = 360° x=1, y=0,** the trig functions values are the same as in θ=0°.
   At θ=360° , x=1, y = 0:

   Sin 360 = y = 0
   Cos 360 = x = 1
   Tan 360 = y/x = 0/1 = 0
   Cot 360 = x/y = 1/0 = undefined
   Sec 360 = 1/x = 1/1 = 1
   Csc 360 = 1/y = 1/0 = undefined

Finding the exact values for the standard Trigonometric Functions for θ= 30°, 45°, 60° :

First we will find the values for θ= 30, and 60 the two complement angles: Using a triangle method as follows: Consider an equal sided triangle where the angles all equal to 60° as shown:

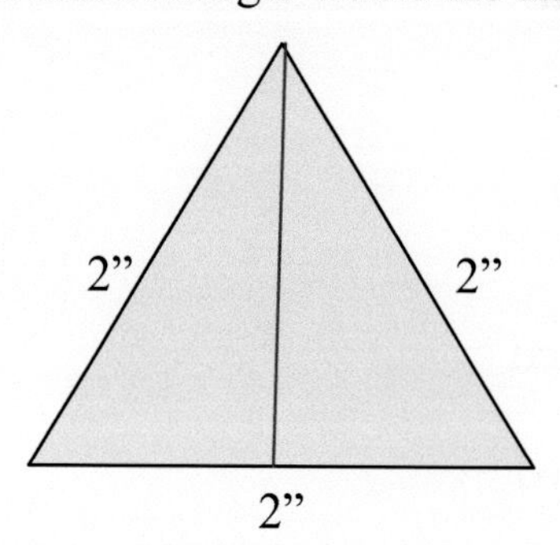

Then divide the triangle in 1/2, so the top angle becomes 30°, and the bottom side becomes 1”:

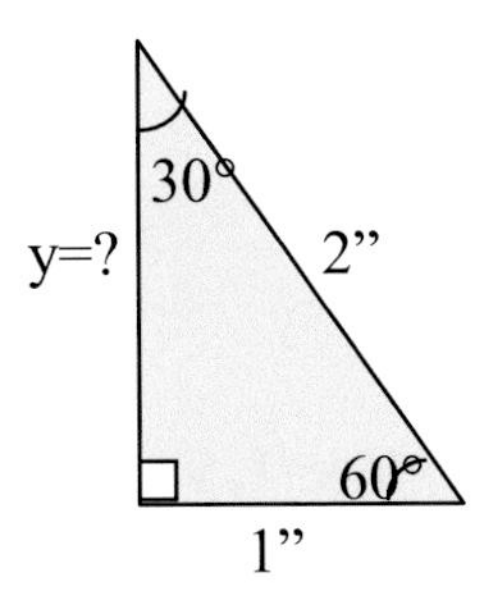

We can find the missing side y by using Pythagorean Theorem: $y = \sqrt{2^2 - 1^2} = \sqrt{3}$
Now we can use this new right Triangle to find all the six-trig function values for the angles 30 and 60°.

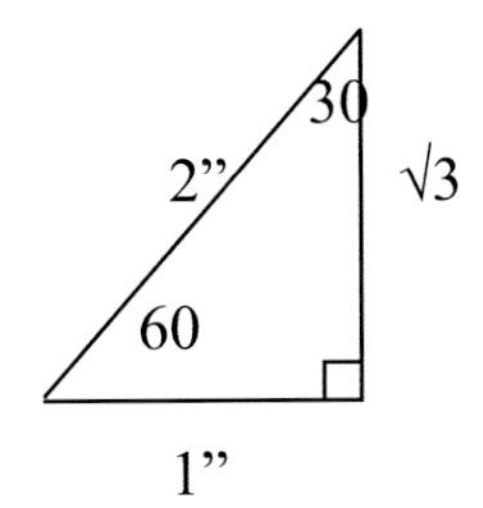

For θ = 30°, x=√3, y= 1, r=2

$$\text{Sin } 30 = \frac{\text{Opposite}}{\text{Hypotenuse}} = \frac{y}{r} = \frac{1}{2}$$

$$\text{Cos } 30 = \frac{\text{Adjacent}}{\text{Hypotenuse}} = \frac{x}{r} = \frac{\sqrt{3}}{2}$$

$$\text{Tan } 30 = \frac{\text{Opposite}}{\text{Adjacent}} = \frac{\text{Sin } \theta}{\text{Cos } \theta} = \frac{y}{x} = \frac{1}{\sqrt{3}}$$

$$\text{Cot } 30 = \frac{\text{Adjacent}}{\text{Opposite}} = \frac{\text{Cos } \theta}{\text{Sin } \theta} = \frac{1}{\text{Tan } \theta} = \frac{x}{y} = \sqrt{3}$$

$$\text{Sec } 30 = \frac{\text{Hypotenuse}}{\text{Adjacent}} = \frac{1}{\text{Cos } \theta} = \frac{r}{x} = \frac{2}{\sqrt{3}}$$

$$\text{Csc } 30 = \frac{\text{Hypotenuse}}{\text{Opposite}} = \frac{1}{\text{Sin } \theta} = \frac{r}{y} = 2$$

For θ = 60°: Using the same triangle x=1, y= √3, r=2:

$$\text{Sin } 60 = \frac{\text{Opposite}}{\text{Hypotenuse}} = \frac{y}{r} = \frac{\sqrt{3}}{2}$$

$$\text{Cos } 60 = \frac{\text{Adjacent}}{\text{Hypotenuse}} = \frac{x}{r} = \frac{1}{2}$$

$$\text{Tan } 60 = \frac{\text{Opposite}}{\text{Adjacent}} = \frac{\text{Sin } \theta}{\text{Cos } \theta} = \frac{y}{x} = \sqrt{3}$$

$$\text{Cot } 60 = \frac{\text{Adjacent}}{\text{Opposite}} = \frac{\text{Cos } \theta}{\text{Sin } \theta} = \frac{1}{\text{Tan } \theta} = \frac{x}{y} = \frac{1}{\sqrt{3}}$$

$$\text{Sec } 60 = \frac{\text{Hypotenuse}}{\text{Adjacent}} = \frac{1}{\text{Cos } \theta} = \frac{r}{x} = \frac{2}{1} = 2$$

$$\text{Csc } 60 = \frac{\text{Hypotenuse}}{\text{Opposite}} = \frac{1}{\text{Sin } \theta} = \frac{r}{y} = \frac{2}{\sqrt{3}}$$

Finding the exact values for the standard Trigonometric Function θ= 45° :

Consider a triangle with two equal sides of 1-unit each and two equal angles of 45°as shown:

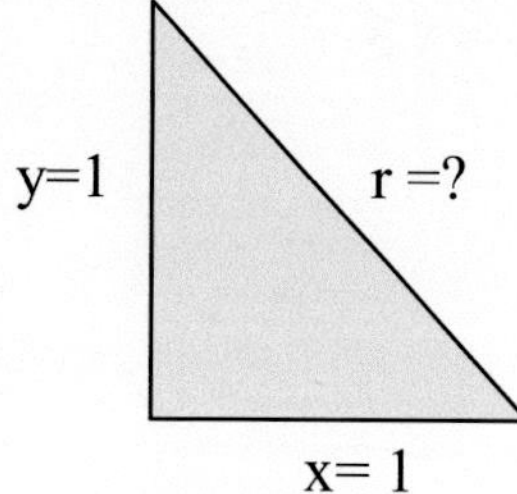

Using Pythagorean Theorem we can find the missing side:

$r = \sqrt{1^2 + 1^2} = \sqrt{2}$

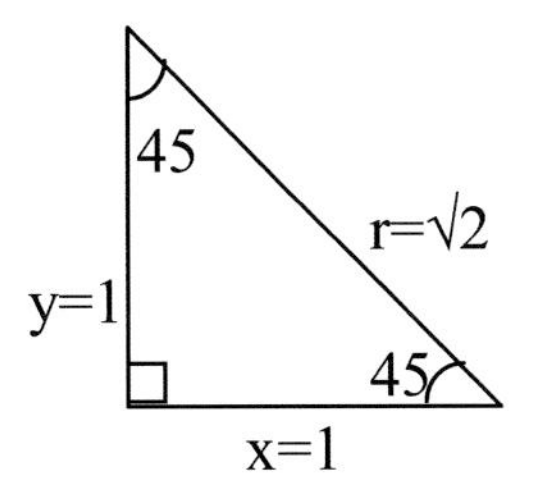

Then the six-Trigonometric Values for θ = 45 are:

$$\text{Sin } 45 = \frac{\text{Opposite}}{\text{Hypotenuse}} = \frac{y}{r} = \frac{1}{\sqrt{2}}$$

$$\text{Cos } 45 = \frac{\text{Adjacent}}{\text{Hypotenuse}} = \frac{x}{r} = \frac{1}{\sqrt{2}}$$

$$\text{Tan } 45 = \frac{\text{Opposite}}{\text{Adjacent}} = \frac{\text{Sin } \theta}{\text{Cos } \theta} = \frac{y}{x} = 1$$

$$\text{Cot } 45 = \frac{\text{Adjacent}}{\text{Opposite}} = \frac{\text{Cos } \theta}{\text{Sin } \theta} = \frac{1}{\text{Tan } \theta} = \frac{x}{y} = 1$$

$$\text{Sec } 45 = \frac{\text{Hypotenuse}}{\text{Adjacent}} = \frac{1}{\text{Cos } \theta} = \frac{r}{x} = \sqrt{2}$$

$$\text{Csc } 45 = \frac{\text{Hypotenuse}}{\text{Opposite}} = \frac{1}{\text{Sin } \theta} = \frac{r}{y} = \sqrt{2}$$

Finding the Exact Values of the Six-Trigonometric Functions for a given point:

Example-5: A point p = (4, 3) is on the terminal side of angle θ. Find the six Trigonometric values for the angle θ.

Solution: From the given point we have: x=4, and y=3. Using Pythagorean Theorem:
$r^2 = x^2 + y^2$, we can find the missing side r (hypotenuse) of the triangle:
$r^2 = 4^2 + 3^2 = 25 \rightarrow$ then $r = \sqrt{25} = 5$
Then the six-Trigonometric Functions of angle θ are:
Sin θ = y/r = 3/5 Cos θ = x/r = 4/5
Tan θ = y/x = ¾ Cot θ = x/y = 4/3
Sec θ = r/x = 5/4 Csc θ = r/y = 5/3

Using Calculator to find the values of non Standard angles

Example - 6: Use TI- to find the following values:

a.Sin 49 b. cos 36 c. tan $\pi/12$ d. cot $\pi/7$ e. Sec 21 Csc 38

Solution:

a. Set the MODE of the TI on degrees, then enter sin49 = .7547095802

b. Cos 36 = .8090169944

c. Set the Mode of TI to Radians, then enter tan $\pi/12$ = .2679491924

d. $\text{Cot}(\pi/7) = 1/\text{Tan}(\pi/7) = 1/\ .4815746188 = 2.076521397$

e. Set the MODE back to degrees, and enter Sec21 = 1/ Cos21 = 1/ .9335804265 = 1.071144994

d. Csc 38 = 1/Sin 38 = 1/ .6156614753 = 1.624269245

Example - 7: In the right triangle find the six-trigonometric functions for:

a. The angle θ on the graph.

b. The angleα on the graph.

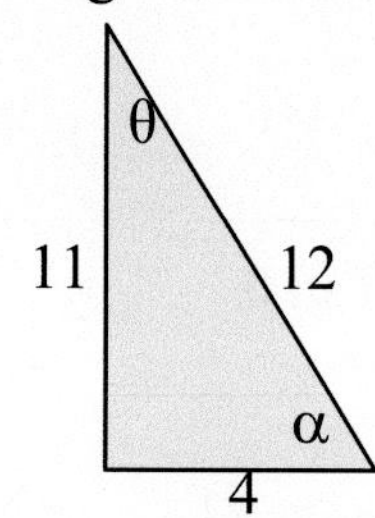

Solution: a. the six trigonometric functions for θ, where Opposite = 4, Adjacent = 11, and Hypotenuse = 12 are:

$$\text{Sin }\theta = \frac{\text{Opposite}}{\text{Hypotenuse}} = \frac{y}{r} = \frac{4}{12}$$

$$\text{Cos }\theta = \frac{\text{Adjacent}}{\text{Hypotenuse}} = \frac{x}{r} = \frac{11}{12}$$

$$\text{Tan }\theta = \frac{\text{Opposite}}{\text{Adjacent}} = \frac{\text{Sin }\theta}{\text{Cos }\theta} = \frac{y}{x} = \frac{4}{11}$$

$$\text{Cot } \theta = \frac{\text{Adjacent}}{\text{Opposite}} = \frac{\text{Cos } \theta}{\text{Sin } \theta} = \frac{1}{\text{Tan } \theta} = \frac{x}{y} = \frac{11}{4}$$

$$\text{Sec } \theta = \frac{\text{Hypotenuse}}{\text{Adjacent}} = \frac{1}{\text{Cos } \theta} = \frac{r}{x} = \frac{12}{11}$$

$$\text{Csc } \theta = \frac{\text{Hypotenuse}}{\text{Opposite}} = \frac{1}{\text{Sin } \theta} = \frac{r}{y} = \frac{11}{12}$$

b. the six trigonometric functions for α, where Opposite = 11, Adjacent = 4, and Hypotenuse = 12 are:

$$\text{Sin } \alpha = \frac{\text{Opposite}}{\text{Hypotenuse}} = \frac{y}{r} = \frac{11}{12}$$

$$\text{Cos } \alpha = \frac{\text{Adjacent}}{\text{Hypotenuse}} = \frac{x}{r} = \frac{4}{12}$$

$$\text{Tan } \alpha = \frac{\text{Opposite}}{\text{Adjacent}} = \frac{\text{Sin } \theta}{\text{Cos } \theta} = \frac{y}{x} = \frac{11}{4}$$

$$\text{Cot } \alpha = \frac{\text{Adjacent}}{\text{Opposite}} = \frac{\text{Cos } \theta}{\text{Sin } \theta} = \frac{1}{\text{Tan } \theta} = \frac{x}{y} = \frac{4}{11}$$

$$\text{Sec } \alpha = \frac{\text{Hypotenuse}}{\text{Adjacent}} = \frac{1}{\text{Cos } \theta} = \frac{r}{x} = \frac{12}{4}$$

$$\text{Csc } \alpha = \frac{\text{Hypotenuse}}{\text{Opposite}} = \frac{1}{\text{Sin } \theta} = \frac{r}{y} = \frac{12}{11}$$

Reciprocal Functions

$$\text{Sec } \theta = \frac{1}{\text{Cos } \theta} \qquad \text{Csc } \theta = \frac{1}{\text{Sin } \theta} \qquad \text{Tan } \theta = \frac{1}{\text{Cot } \theta} \qquad \text{Cot } \theta = \frac{1}{\text{Tan } \theta}$$

Note: The value of the trigonometric function for any angle depends on the measure of the angle, and not on the size of the triangle.

5.3 Solving Right Triangles

The right Triangle is a triangle with one angle =90°. To solve a right triangle means to find the missing angles and sides.

Example-8: Solve the right triangle with r=3.45m, α=17.8°.

Solution: To solve this triangle, means to find: angle θ, and the Two sides: x and y: To find the angle θ we use the following Triangle theorem:

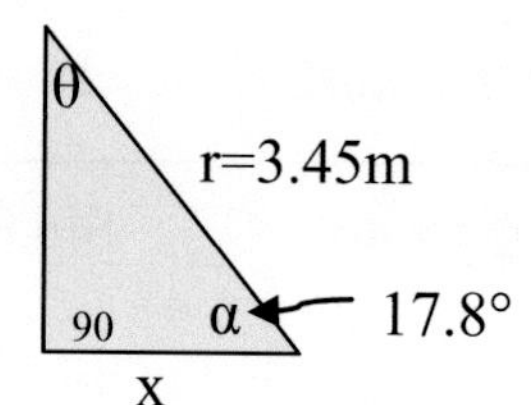

Theorem: Total Angles of Any Triangle = 180°

that is: $\theta + \alpha + 90° = 180$

$\theta + 17.8 + 90 = 180$ solving for θ, we get: θ= 72.2°

Now to find the missing sides, we use Pythagorean Theorem for right triangle:

Pythagorean Theorem for Right Triangle: $r^2 = x^2 + y^2$

Since two sides are missing then we cannot use the above theorem, because of 2-missings, then we have to use the Trigonometric functions to find them as follows:

$$\text{Sin } \alpha = \frac{\text{Opposite}}{\text{Hypotenuse}} = \frac{y}{r} = \frac{y}{3.45}$$

Solving for y gives: y = 3.45 Sin (17.8) = 1.05 m.

Now we can apply Pythagorean Theorem to find the x-side:

$x = \sqrt{r^2 - y^2} = \sqrt{(3.45)^2 - (1.05)^2} = 3.28$ m.

Or we can use Trigonometric Function to solve as follows:

$$\text{Cos } \alpha = \frac{\text{Adjacent}}{\text{Hypotenuse}} = \frac{x}{r} = \frac{x}{3.45}$$

Solving for x gives: x = 3.45 Cos (17.8) = 3.28 m

Theorems:

1. If any two sides of a right triangle are known, then it is possible to solve for the remaining sides and the three angles.
2. If θ and α are the acute angles of a right triangle, then:

   Sin θ = Sinα, Sec θ = Cscα

   Tan θ = Cotα, Csc θ = Secα

   Cot θ = Tanα,

Example-9: Find the angle between the line $y = 0.5x + 3$ and the x-axis, in degrees to one decimal place.
Solution: Graph the equation to see the angle:

y=0.5x+3

α

x-axis

Here we use the slope definition:

$$\text{Slope} = m = \frac{\text{Rise}}{\text{Run}} = \frac{y}{x} = 0.5$$

$$\text{But Tan } \alpha = \frac{\text{Opposite}}{\text{Adjacent}} = \frac{y}{x} = 0.5$$

Then: $\text{Tan } \alpha = 0.5 \rightarrow \alpha = \text{Tan}^{-1} 0.5 = 26.6°$

5.4 Trigonometric Function's Properties

Domain and Range of Trigonometric Functions

To understand the domain of all the trigonometric functions, it's better to look at the graph of each function. Since the domain represents all the x-values, and the range represents all the y-values, then to find the **Domain** function from its graph, picture if the graph is crushed on x-axis, in the same way to find the **Range** of the function graph, picture the function crushed on y-axis.

$y = \text{Sin}(x)$

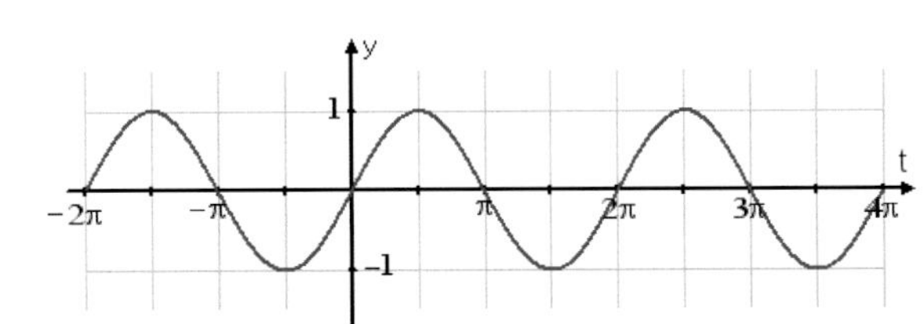

The Domain of $y = \text{Sin}(x)$ is all the real numbers or $\{-\infty < x < \infty\}$, and
The range of $y = \text{Sin}(x)$ is $\{-1 \leq y \leq 1\}$ or [-1, 1].
The graph is symmetric with respect to the origin.

$y = \text{Cos}(x)$

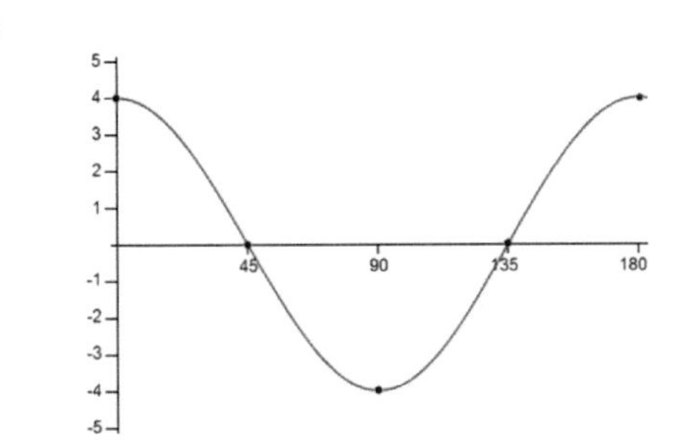

The Domain of y = Cos(x) is all the real numbers $\{-\infty < x < \infty\}$, and
The range of y= Cos(x) is $\{-1 \leq y \leq 1\}$ or [-1, 1].
The graph is symmetric with respect to the y-axis

y=Tan(x)

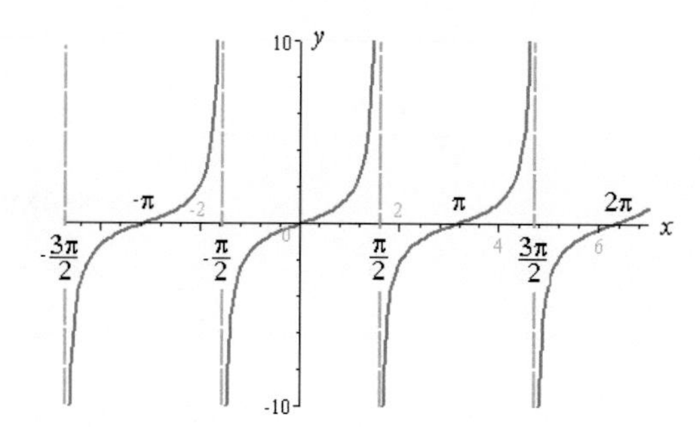

The Domain of y = Tan(x) is all the real numbers EXCEPT $\pi/2 + k\pi$, k is an integer.
The range of y= Tan(x) is all the real numbers or $\{-\infty < x < \infty\}$.
The graph is symmetric with respect to the origin.

y=Cot(x)

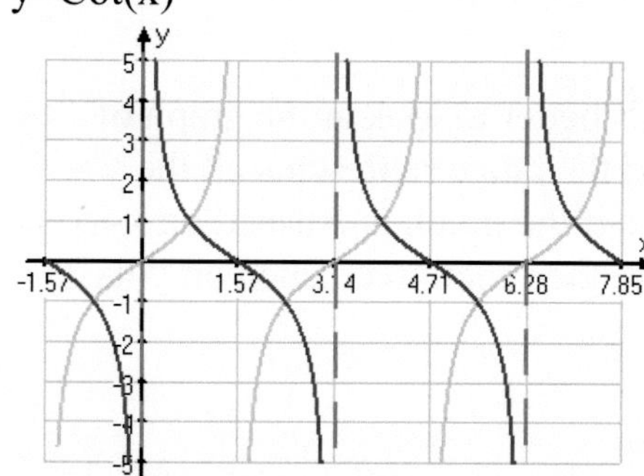

The Domain of y = Cot(x) is all the real numbers EXCEPT $k\pi$, k is an integer.
The range of y= Cot(x) is all the real numbers or $\{-\infty < x < \infty\}$.
The graph is symmetric with respect to the origin.

y=Sec(x)

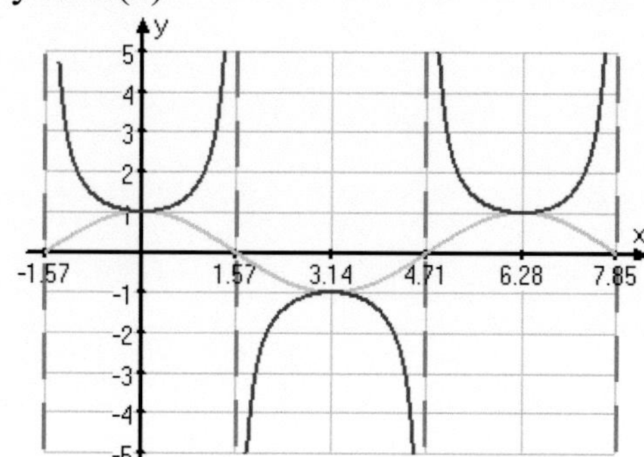

The Domain of y = Sec x is all the real numbers EXCEPT $\pi/2 + k\pi$, k is an integer
The range of y= Sec x is all the real numbers such that $y \leq -$, or $y \geq 1$.
The graph is symmetric with respect to y-axis.

y=Csc(x)

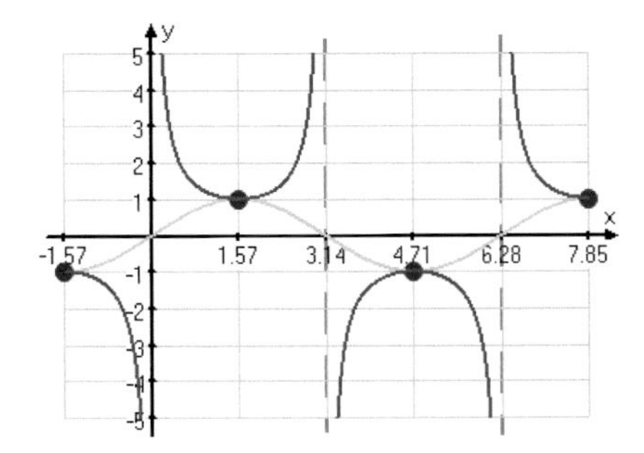

The Domain of y = Csc(x) is all the real numbers EXCEPT $k\pi$, k is an integer.
The range of y= Csc(x) is all the real numbers such that $y \leq -$, or $y \geq 1$.
The graph is symmetric with respect to the origin.

Graph of Sine and Cosine Functions

Here we will graph the sine and cosine functions using tables and unit circle, this can be applied to any trigonometric function.

Example-10: Graph the function y=sin θ in the interval [0,360]:

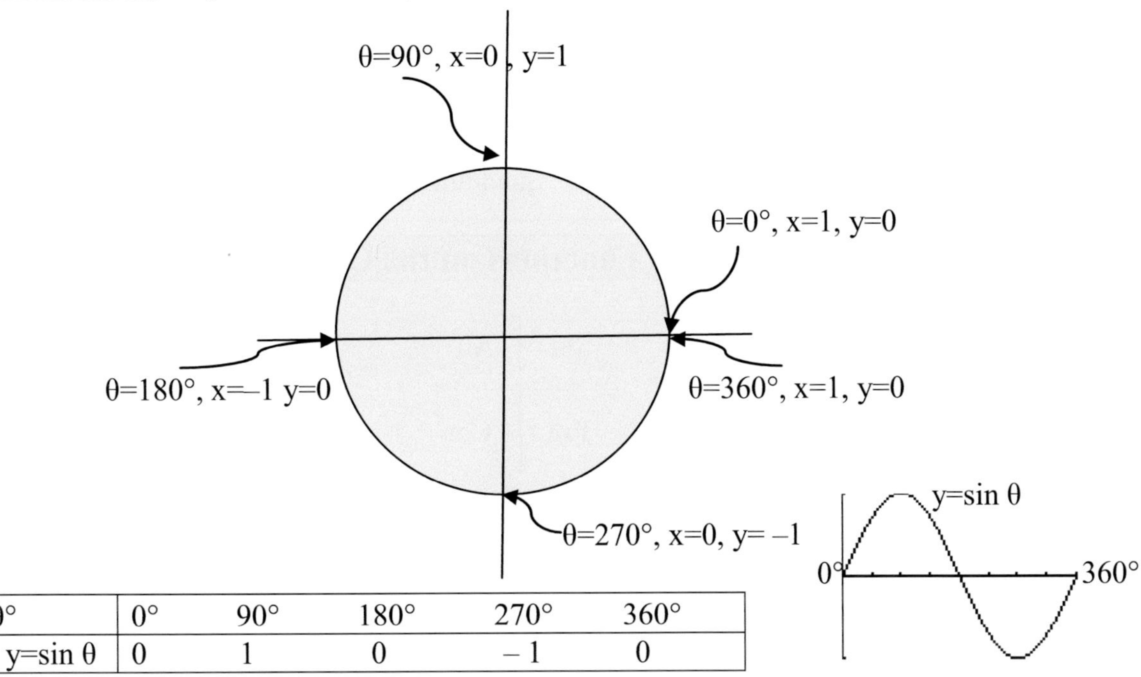

θ°	0°	90°	180°	270°	360°
y=sin θ	0	1	0	– 1	0

Example-11: Graph the function y= cos θ in the interval [0,360]:

θ°	0°	90°	180°	270°	360°
y= cos θ	1	0	– 1	0	1

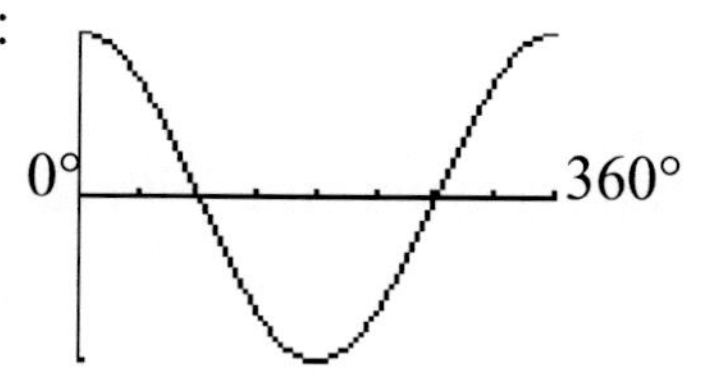

The Periodic Properties

$\text{Sin}(\theta + 2\pi k) = \sin \theta$ $\quad$ $\text{Cos}(\theta + 2\pi k) = \text{Cos } \theta$ $\quad$ $\text{Tan}(\theta + \pi k) = \tan \theta$

$\text{Csc}(\theta + 2\pi k) = \text{Csc } \theta$ $\quad$ $\text{Sec}(\theta + 2\pi k) = \text{Sec } \theta$ $\quad$ $\text{Cot}(\theta + \pi k) = \text{Cot } \theta$

Where k is an integer

Example- 12: If sin θ = 0.4, then the value of:
$\text{Sin }(\theta+4\pi) + \text{Sin } \theta + \text{Sin}(\theta+\pi) = \text{Sin } \theta + \text{Sin } \theta + \text{Sin } \theta = 3 \text{ Sin } \theta = 3(0.4) = 1.2$

Example-13: Using Periodic Property, Find the exact values of:
a. Sin (15π/4) b. Cos (3π) c. Tan (7π/4)

Solution:

a. Sin (15π/4) = Sin (15 x 45°) This falls on the 4th quadrant = sin45 = - 1/√2

b. Cos (3π) = Cot (3 x 180°) This falls on the second quadrant on the x-axis
= Cot (180°) = -1

c. Tan (7π/4) = Tan (7 x 45°) Falls on the 4th quadrant = Tan 45° = -1

The Positive Signs of Trigonometric Functions on the Quadrant

Q_{II}	Q_{I}
Q_{III}	Q_{IV}

(-x,y)	(x,y)
(-x,-y)	(x,-y)

Sin	All
Tan	Cos

Quadrant-I: All the trigonometric functions are plosive in the first quadrant.
Quadrant-II: Sine and its reciprocal (Csc) are positive in the second quadrant.
Quadrant-III: Tan and its reciprocal (Cot) are positive on the third quadrant
Quadrant-IV: Cos and its reciprocal (sec) are positive on the fourth quadrant.

Example-14: Given that Tan α = – 2/3 and θ is in the third quadrant. Find the other trigonometric function values.

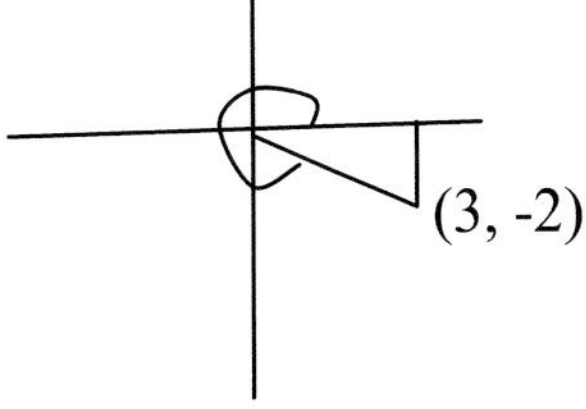

Solution: First we graph the triangle:
From the given angle we have: y= –2, and x=3

$$\text{Tan } \alpha = -2/3 = \frac{-y}{x}$$

We have to find the hypotenuse using Pythagorean Theorem: $r = \sqrt{x^2 + y^2} = \sqrt{13}$
Then the rest of the trigonometric functions are:

$$\text{Sin } \alpha = \frac{\text{Opposite}}{\text{Hypotenuse}} = \frac{y}{r} = \frac{-2}{\sqrt{13}}$$

$$\text{Cos } \alpha = \frac{\text{Adjacent}}{\text{Hypotenuse}} = \frac{x}{r} = \frac{3}{\sqrt{13}}$$

$$\text{Cot } \alpha = \frac{\text{Adjacent}}{\text{Opposite}} = \frac{\text{Cos } \theta}{\text{Sin } \theta} = \frac{1}{\text{Tan } \theta} = \frac{x}{y} = \frac{3}{-2}$$

$$\text{Sec } \alpha = \frac{\text{Hypotenuse}}{\text{Adjacent}} = \frac{1}{\text{Cos } \theta} = \frac{r}{x} = \frac{\sqrt{13}}{3}$$

$$\text{Csc } \alpha = \frac{\text{Hypotenuse}}{\text{Opposite}} = \frac{1}{\text{Sin } \theta} = \frac{r}{y} = \frac{\sqrt{13}}{-2}$$

Even Odd Properties

For a given function y = f(x) when replacing x with –x(x→-x):
1. If f(- x) = f(x) → then the function is said to be even function.
2. If f(-x) = -f(x) then the function is said to be odd function.

Sin(– θ) = – Sin θ (odd)	Cos(– θ) = Cos θ (even)	Tan(– θ) = –tan θ(odd
Csc(– θ) = – Csc θ (odd)	Sec(– θ) = Sec θ (even)	Cot(– θ) = – Cot θ (odd)

Example - 15: Using Odd-Even Property, find the exact values of:
a. Sin (–90) b. Cos (-45) c. Cot (-60) d. Tot (30)

Solution:
a. Sin (-90) = - Sin 90 = -1
b. Cos (-45) = Cos 45 = $1/\sqrt{2}$
c. Cot (-60) = – Cot 60 = $\sqrt{3}$
d. Tan (–30) = – Tan 30 = $1/\sqrt{3}$

5.5 Trigonometric Identities

Reciprocal Identities

$$\text{Sec } \theta = \frac{1}{\text{Cos } \theta} \qquad \text{Csc } \theta = \frac{1}{\text{Sin } \theta} \qquad \text{Tan } \theta = \frac{1}{\text{Cot } \theta} \qquad \text{Cot } \theta = \frac{1}{\text{Tan } \theta}$$

Quotient Identities

$$\text{Tan } \theta = \frac{\text{Sin } \theta}{\text{Cos } \theta} \qquad \text{Cot } \theta = \frac{\text{Cos } \theta}{\text{Sin } \theta}$$

Pythagorean Identities

Consider a right triangle inscribed inside a unit circle with r=1, as shown in the graph:

Using Pythagorean Theorem: we get:
$x^2 + y^2 = 1$ --------- (1)
But in unit circle we have found that:
Sin θ = y, and Cos θ=x
Substituting in the above relation (1) we gat:
$\text{Sin}^2\ \theta + \text{Cos}^2\ \theta = 1$ this is the first trigonometric Identity.
Working algebraically on (1) we can come up with different Identities as follows:

Divide equation (1) by x^2 gives: $1 + \frac{y^2}{x^2} = \frac{1}{x^2}$

Or → $1 + (y/x)^2 = (1/x)^2$ --- (2)
But y/x = Tan θ , and 1/x = sec θ, then equation (2) is:

$1 + \text{Tan}^2\ \theta = \text{Sec}^2\ \theta$, a second Identity. In the same manner if we divide equation (1) by y^2 gives a third Identity: $\text{Cot}^{2\ \theta} + 1 = \text{Csc}^2\ \theta$, then the Pythagorean Identities and their equivalents are:

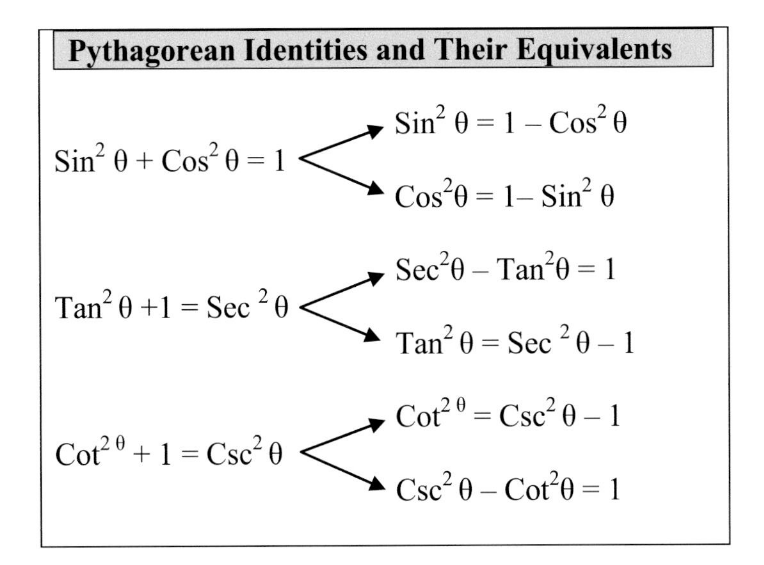

Complementary Angles: Are the angles that adds up to $90°(\pi/2)$

$\theta = 0°$ Complement of $90°$	**$\theta = 30°$ Complement of $60°$**
Sin 0 = Cos 90	Sin 30 = Cos 60
Cos 0 = Sin 90	Cos 30 = Sin 60
Tan 0 = Cot 90	Tan 30 = Cot 60
Cot 0 = Cot 90	Cot 30 = Cot 60
Sec 0 = Csc 90	Sec 30 = Csc 60
Csc 0 = Sec 90	Csc 30 = Sec 60

Supplementary Angles: Are angles that adds up to $180°(2\pi)$

Reference Angles (α): If angle θ lies in a quadrant, then its reference angle is α, which is the angle between the x-axis and the terminal as shown below:

A. If the terminal is on the first quadrant with angle θ, then the reference angle $\alpha = \theta$.
B. If the terminal is in the second quadrant with angle θ, the reference angle α is the Supplement angle to θ or $\alpha = 180 - \theta$
C. If the terminal is on the third quadrant with angle θ, then the reference angle $\alpha = \theta - 180$.
D. If the terminal is on the fourth quadrant with angle θ, then the reference angle $\alpha = \theta - 360$.

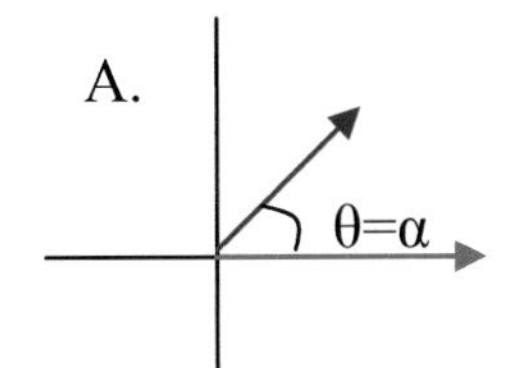

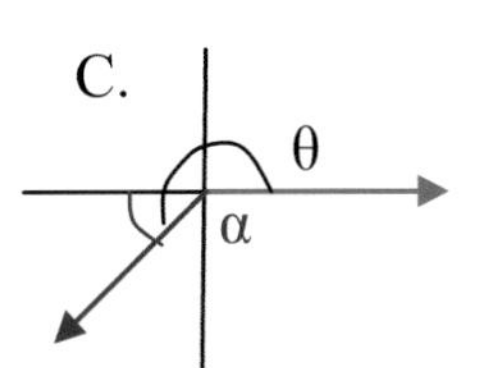

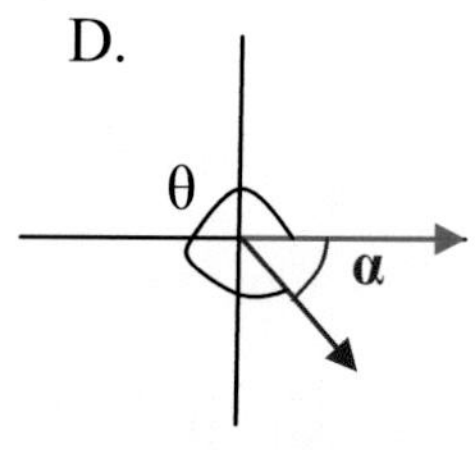

Example-16	Finding Reference Angles

Find the reference angle α for the given positive angles θ:

A. In the first quadrant θ= 30° → the reference angle α = 30°
B. In the second quadrant θ=145° → the reference angle α = 180° – 145° = 35°
C. In the third quadrant θ = 265° → the reverence angle α = 265° – 180° =85°
D. In the 4th quadrant the angle θ = 300° → the reference angle α = 360° – 300° = 60°

Find the reference angle α for the given negative angles θ:

A. For θ= –30 → the reference angle α = +30

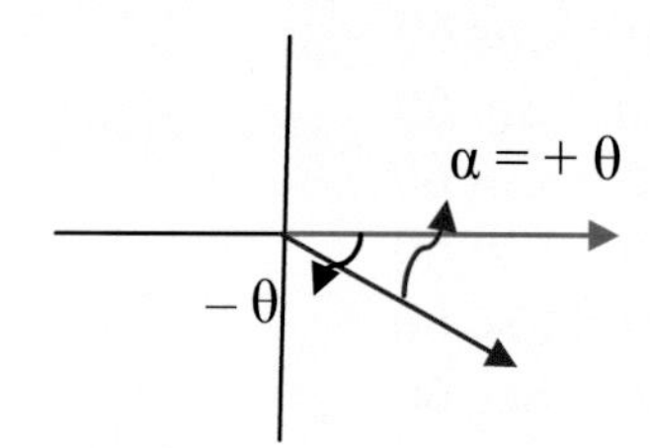

B. θ = – 130° , the reference angle α = 180° – 130° = 50°

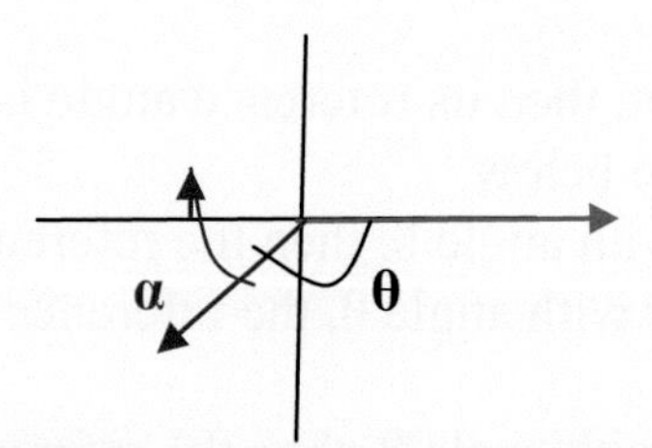

Reference Angles

Sin(θ) = ±Sin α	Cos(θ) = ±Cos α	Tan(θ) = ±Tan α
Csc(θ) = ±Csc α	Sec(θ) = ±Sec α	Cot(θ) = ±Cot α

<u>Example-17:</u> Using the reference angles, find the following exact values:
a. Sin 150   b. Cos 135   c. Tan (–2π/3)   d. Sec (5π/6)

Solution:

a. Sin 150 = Sin($\pi/2 + 60$) the reference angle in the second quadrant

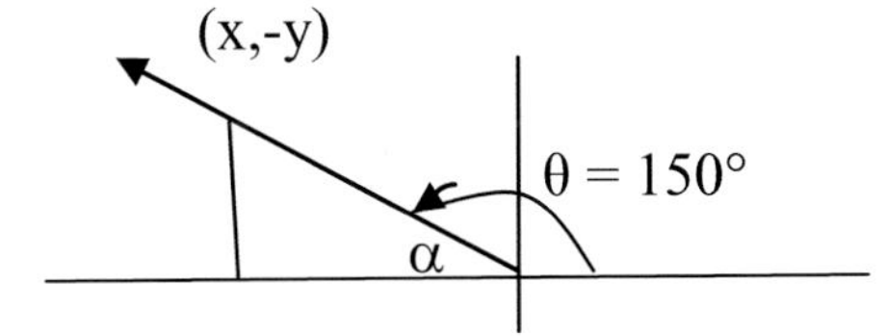

The reference angle $\alpha = 180 - 150 = 30°$, then Sin 150 = Sin 30 = 1/2

b. Cos 135 = Cos (90 + 45), the reference angle falls in the second quadrant α=45°

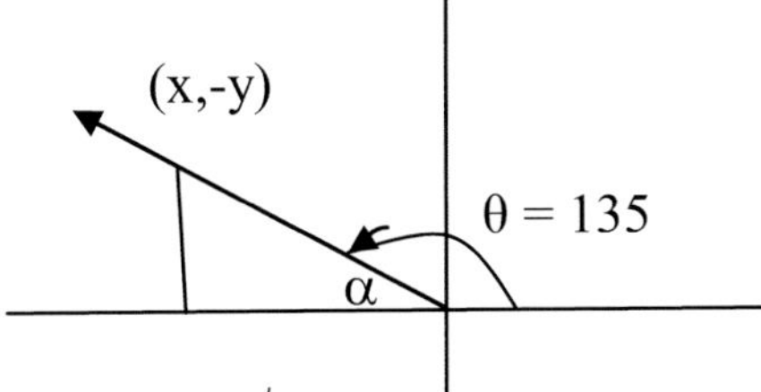

Then Cos 135 = Cos 45 = $-1/\sqrt{2}$

a. Tan $(-2\pi/3)$ = Tan (120) = Tan(90 + 30) Its reference angle α=60 on the second quadrant
   Then Tan $(-2\pi/3)$ = Tan 60 = $\sqrt{3}$

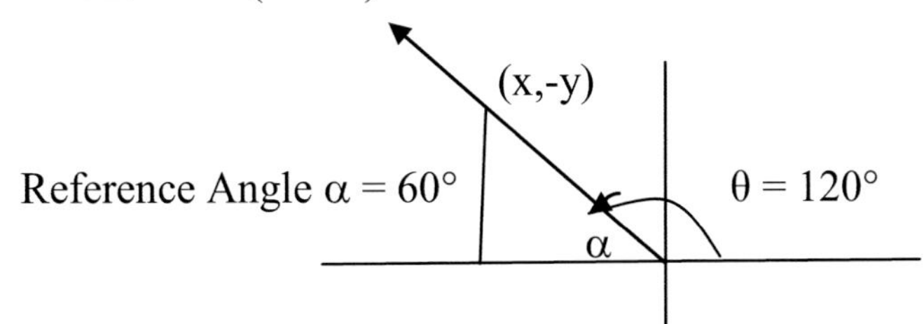

5.6 Trigonometric Applications

Trigonometric equations can be used to estimates lengths, or distance, of shapes that cannot be reached easily.

Example: Anna wants to find how tall the tree in her backyard is? she estimated the angle of elevation of the tree's peak to be 80°. She walks off 40 feet from the tree, then use trigonometric of right triangle to find the height of the tree as follows:

$$\text{Tan } 80° = \frac{\text{Opposite}}{\text{Adjacent}} = \frac{H}{40 \text{ ft}}$$

Then → H = 40 Tan80 ≈ 226.9 ft

H

80°

40 ft

Chapter - 5 Exercise

1. In the triangle below, $c = 215$ ft. Solve the triangle for sides a, and b side.

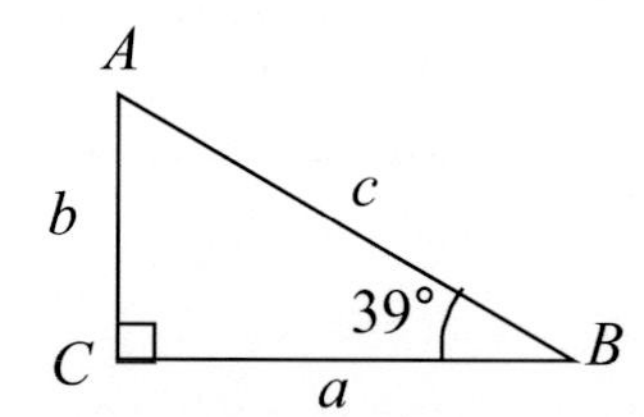

2. Find the measure of angle θ to the nearest tenth of a degree.

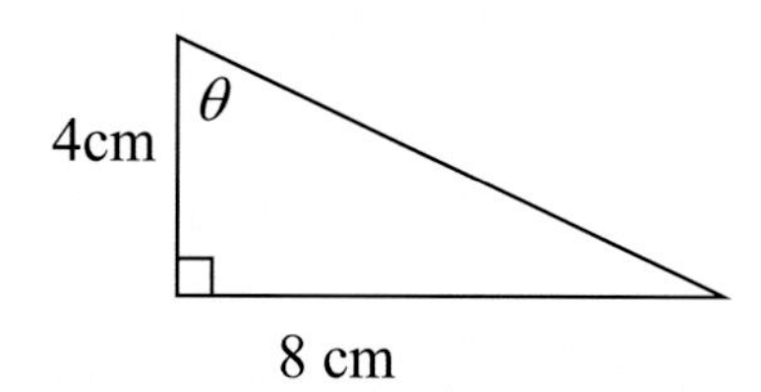

3. Find the measure of angle α to the nearest tenth of a degree.

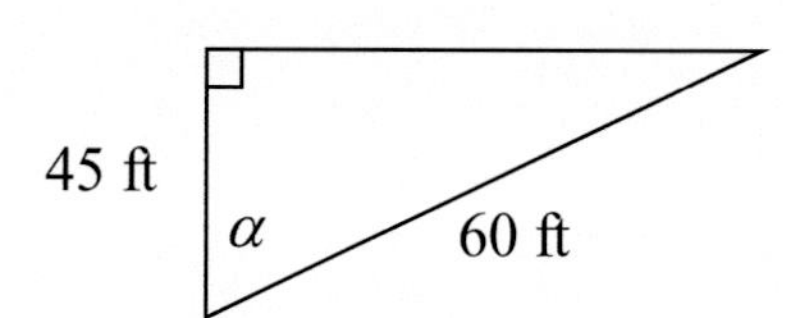

4. Find the degree measure to one decimal place of the acute angle between the given line and the x-axis.

   $$y = 2/3x + 3$$

5. The point $\left(-\frac{1}{6}, -\frac{\sqrt{2}}{3}\right)$ is on the terminal side of angle θ with θ in standard position. Find the value of the $\cos(\theta)$.

6. Find the reference angle(α) for $\theta = 240°$

7. Find the reference angle $\theta = -7\pi/4$

8. Find the value of cot θ if $\sin\theta = -\frac{19\sqrt{6}}{5}$ and $\cos\theta > 0$

9. A person is standing 150 ft from a building. The angle to the top of building relative to the horizon is 54.7. How tall is the building?

10. The point (4, -12) is on the terminal side of the angle θ with θ in standard position. Find the value of the 6-trig. Functions:

Chapter - 5 Test

1. Find the measure of angle α to the nearest tenth of a degree.

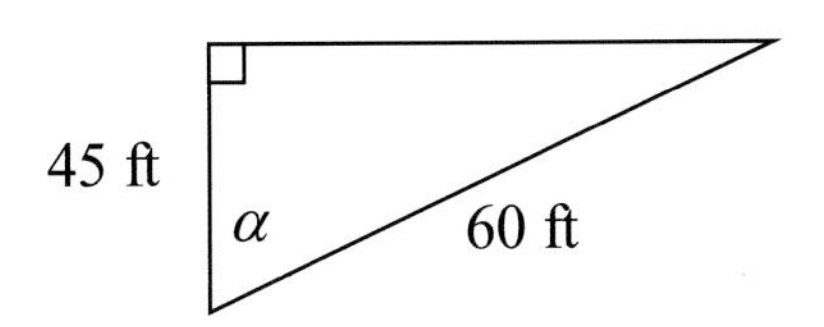

2. person is standing 250 ft from a building . The angle to the top of building relative to the horizon is 53.8. How tall is the building?

3. The point (4, -12) is on the terminal side of the angle θ with θ in standard position. Find the value of the 6-trig. Functions:

4. Find the reference angle for θ= 200°, 350°

6. Trigonometric Identities

Chapter – 6
Trigonometric Identities, Inverse Functions And Equations

Objectives:
6.1 Sum and Difference Identities
6.2 Double-Angles, and Half-Angle
6.3 Proving Trigonometric Identities
6.4 Inverse of the Trigonometric Functions
6.5 Solving Trigonometric Equations

6.1 Sum and Difference Identities

Sum and Difference Formulas for Sine Function

$\text{Sin}(\alpha + \beta) = \text{Sin}\,\alpha\,\text{Cos}\,\beta + \text{Cos}\alpha\,\text{Sin}\,\beta$ (1)

If $\alpha = \beta$

$\text{Sin}(\alpha + \alpha) = \text{Sin}(2\alpha) = \text{Sin}\,\alpha\,\text{Cos}\,\alpha + \text{Cos}\alpha\,\text{Sin}\,\alpha$

$= 2\,\text{Sin}\,\alpha\,\text{Cos}\,\alpha$

$\text{Sin}(\alpha - \beta) = \text{Sin}\,\alpha\,\text{Cos}\,\beta - \text{Cos}\alpha\,\text{Sin}\,\beta$ (2)

$\text{Sin}(\pi/2 - \alpha) = \text{Sin}\,\alpha$ (3)

Sum and Difference Formulas for Cosine Function

$\text{Cos}(\alpha + \beta) = \text{Cos}\alpha\,\text{Cos}\,\beta - \text{Sin}\alpha\,\text{Sin}\,\beta$ (4)

If $\alpha = \beta$

$\text{Cos}(\alpha + \alpha) = \text{Cos}(2\alpha) = \text{Cos}\alpha\,\text{Cos}\,\alpha - \text{Sin}\alpha\,\text{Sin}\,\alpha$

$= \text{Cos}^2\alpha - \text{Sin}^2\alpha$

$\text{Cos}(\alpha - \beta) = \text{Cos}\alpha\,\text{Cos}\,\beta + \text{Sin}\alpha\,\text{Sin}\,\beta$ (5)

$\text{Cos}(\pi/2 - \alpha) = \text{Sin}\alpha$ (6)

Sum and Difference for Tangent Function

$$\text{Tan}(\alpha + \beta) = \frac{\text{Tan}\,\alpha + \text{Tan}\,\beta}{1 - \text{Tan}\alpha\text{Tan}\,\beta} \quad (7)$$

$$\text{Tan}(\alpha - \beta) = \frac{\text{Tan}\alpha - \text{Tan}\,\beta}{1 + \text{Tan}\alpha\text{Tan}\,\beta} \quad (8)$$

Example-1: Find the exact value of cos (75)

Solution: Sin (75) = Sin (45 + 30)
= Sin 45 Cos 30 + Cos 45 Sin 30
= $(1/\sqrt{2})(\sqrt{3}/2) + (1/\sqrt{2})(1/2)$
= $\sqrt{3} + 1 / 2\sqrt{2} = 1/2\sqrt{2}(\sqrt{3} + 1)$

Example-2: Find the exact value of : Sin 40 Cos20 + Cos40 Sin20

Solution: Let $\alpha = 40$, and $\beta=20$
$\sin(\alpha+\beta) = \sin\alpha \cos\beta + \cos\alpha \sin\beta$
Then → Sin 40 Cos20 + Cos40 Sin20 = Sin (40+20) = Sin 60 = ½

Exapmle-3: Find the exact value of Cos ($7\pi/12$)

Solution: Cos ($7\pi/12$) = Cos ($3\pi/12 + 4\pi/12$)
= Cos ($\pi/4 + \pi/3$)
= Cos $\pi/4$ Cos $\pi/3$ – Sin $\pi/4$ Sin $\pi/3$
= Cos 45 Cos 60 – Sin 45 Sin 60
= $(1/\sqrt{2})(1/2) - (1/\sqrt{2})(\sqrt{3}/2)$
= $(1/2\sqrt{2})(1 - \sqrt{3})$

Example-4: Given Sinα = 3/4 for $\pi/2 < \alpha < \pi$, and Sin β = -1/2 for $\pi < \beta < 3\pi/2$.
Find the exact value of: a. Cos α b. Cos β c. Cos ($\alpha+\beta$) d. Sin ($\alpha + \beta$)

Solution: From the given information α is in the second quadrant. We graph and find the rest of the trig functions for α
We need to find the missing value x
Using Pythagorean Theorem:

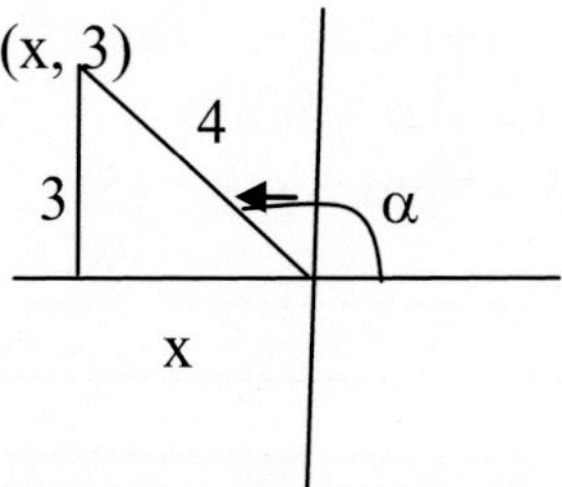

$x^2 + 9 = 16$, $x<0$
Solving for x gives: $x = -\sqrt{7}$ in the second quadrant.
Then → a. Cos α = x/r = $-\sqrt{7}/4$
Also from the given information β is in the third quadrant.

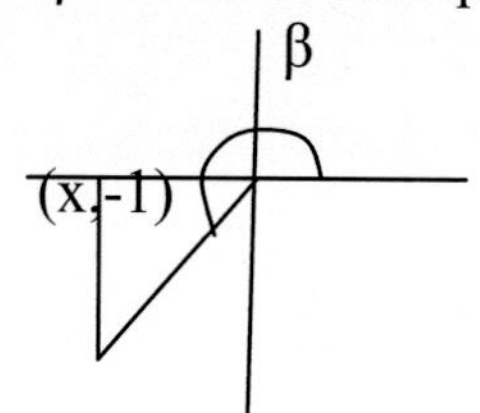

Using Pythagorean Theorem we find $x = -\sqrt{3}$.

→ b. $\cos\beta = x/r = -\sqrt{3}/4$

→ c. $\cos(\alpha+\beta) = \cos\alpha \cos\beta - \sin\alpha \sin\beta$

$$= (-\sqrt{7}/4)(-\sqrt{3}/4) - (3/4)(-1/2)$$
$$= \sqrt{21}/16 + 3/8 = (1/16)(\sqrt{21} + 6)$$

d. $\sin(\alpha+\beta) = \sin\alpha \cos\beta + \cos\alpha \sin\beta$

$$= 3/4(-\sqrt{3}/4) + (-\sqrt{7}/4)(-1/2)$$
$$= -3\sqrt{3}/16 + \sqrt{7}/8$$
$$= (-3\sqrt{3} + 2\sqrt{7})/16$$

6.2 Co functions, Double-Angles, and Half-Angles

Co functions Identity

$\sin(\pi/2 - \theta) = \cos\theta$ $\quad$ $\cos(\pi/2 - \theta) = \sin\theta$

$\tan(\pi/2 - \theta) = \cot\theta$ $\quad$ $\cot(\pi/2 - \theta) = \tan\theta$

$\sec(\pi/2 - \theta) = \csc\theta$ $\quad$ $\csc(\pi/2 - \theta) = \sec\theta$

Example-5: Find the equivalent expression for the folioing Identities:

a. $\cos(\pi/2 + \theta)$ $\qquad$ $\sin(\pi/2 - \theta)$

Solution: **a.** $\cos(\pi/2 + \theta) = \cos\pi/2 \cos\theta - \sin\pi/2 \sin\theta$

$$= (0)\cos\theta - (1)\sin\theta = -\sin\theta$$

b. $\sin(\pi/2 - \theta) = \sin(\pi/2)\cos\theta - \cos(\pi/2)\sin\theta = \cos\theta$

Co functions Identity for Sine and Cosine

$\sin(\pi/2 \pm \theta) = \pm\cos\theta$ $\qquad$ $\cos(\pi/2 - \theta) = +\sin\theta$ $\qquad$ $\cos(\pi/2 + \theta) = -\sin\theta$

Double Angle Identities

$\sin 2\theta = 2\sin\theta\cos\theta$

$\cos 2\theta = \cos^2\theta - \sin^2\theta$ → $\cos 2\theta = 1 - 2\sin^2\theta$; → $\cos 2\theta = 2\cos^2\theta - 1$

$\tan 2\theta = 2\tan\theta / 1 - \tan^2\theta$

Square Identities	
$\text{Sin}^2 \theta = \dfrac{1 - \text{Cos}2\theta}{2}$	$\text{Sin}^2 \theta/2 = \dfrac{1 - \text{Cos}\,\theta}{2}$
$\text{Cos}^2\theta = \dfrac{1 + \text{Cos}2\theta}{2}$	$\text{Cos}^2 \theta/2 = \dfrac{1 + \text{Cos}\,\theta}{2}$
$\text{Tan}^2\theta = \dfrac{1 - \text{Cos}\,2\theta}{1 + \text{Cos}\,2\theta}$	$\text{Tan}^2 \theta/2 = \dfrac{1 - \text{Cos}\,\theta}{1 + \text{Cos}\,\theta}$

Example-6: Prove the Identity $\text{Sin}\,2\theta = 2\,\text{Sin}\,\theta\,\text{Cos}\,\theta$
Solution: $\text{Sin}\,2\theta = \text{Sin}\,(\theta + \theta) = \text{Sin}\,\theta\,\text{Cos}\,\theta + \text{Cos}\,\theta\,\text{Sin}\,\theta = 2\,\text{Sin}\,\theta\,\text{Cos}\,\theta$

Example-7: Prove the Identity: $\text{Cos}\,2\theta = \text{Cos}^2\,\theta - \text{Sin}^2\,\theta$
Solution: $\text{Cos}\,2\theta = \text{Cos}\,(\theta + \theta) = \text{Cos}\,\theta\,\text{Cos}\,\theta - \text{Sin}\,\theta\,\text{Sin}\,\theta$
$= \text{Cos}^2\,\theta - \text{Sin}^2\,\theta$

The last Identity can be written also as:
$\text{Cos}^2\theta - \text{Sin}^2\,\theta = 1 - \text{Sin}^2\theta - \text{Sin}^2\theta$
$= 1 - 2\,\text{Sin}^2\theta$
Then $\text{Cos}\,2\theta = 1 - 2\,\text{Sin}^2\,\theta \rightarrow \text{Sin}^2\,\theta = (1 - \cos 2\theta) / 2$
In a similar way we can find $\text{Cos}^2\theta = (1 + \text{Cos}2\theta) / 2$

6.3 Proving Identities

To prove an Identity is to show that the right side is equal to the left side. So we simplify the right side, and the left side until it is the same expression on both sides.

Example-8: Prove $\text{Sec}\,\theta\,.\,\text{Sin}\,\theta = \text{Tan}\,\theta$

Solution:
R.S $\rightarrow$ $\text{Tan}\,\theta = \text{Sin}\,\theta / \text{Cos}\,\theta$
L.S $\rightarrow$ $\text{Sec}\,\theta\,.\,\text{Sin}\,\theta = 1/\,\text{Cos}\,\theta\,.\,\text{Sin}\,\theta = \text{Sin}\,\theta / \text{Cos}\,\theta = \text{Tan}\,\theta$
Then R.S = L.S $\rightarrow$ then this is an Identity.

Example-9: Prove $\text{Sin}\,\theta\,\text{Csc}\,\theta - \text{Cos}^2\,\theta = \text{Sin}^2\,\theta$

Solution: L.S $\rightarrow$ $\text{Sin}\,\theta\,\text{Csc}\,\theta - \text{Cos}^2\,\theta = \text{Sin}\,\theta\,(1/\text{Sin}\,\theta) - \text{Cos}^2\,\theta$

$= 1 - \text{Cos}^2\,\theta = \text{Sin}^2\,\theta = \text{R.S}$

Example-10: $\text{Tan}\,\theta + \cot\,\theta - \text{Sec}\,\theta\,\text{Csc}\,\theta = 0$

Solution:

L.S → $\tan\theta + \cot\theta - \sec\theta \csc\theta = \frac{\sin\theta}{\cos\theta} + \frac{\cos\theta}{\sin\theta} - \frac{1}{\cos\theta} \cdot \frac{1}{\sin\theta}$

$$= \frac{\sin^2\theta + \cos^2\theta - 1}{\cos\theta \sin\theta} = \frac{1 - 1}{\cos\theta \sin\theta} = 0$$

Example-11: Prove $\cos(\alpha + \beta) + \cos(\alpha - \beta) = 2\cos\alpha \cos\beta$

Solution:

L.S → $\cos(\alpha + \beta) + \cos(\alpha - \beta) = \cos\alpha \cos\beta - \sin\alpha \sin\beta + \cos\alpha \cos\beta + \sin\alpha \sin\beta$
$= 2\cos\alpha \cos\beta =$ R.S

Example-12: Prove $\sin(\pi + \alpha) = -\sin\alpha$

Solution: L.S. → $\sin(\pi + \alpha) = \sin\pi \cos\alpha + \cos\pi \sin\alpha$
$= (0)\cos\alpha + (-1)\sin\alpha = -\sin\alpha =$ R.S

Example-13: Prove $\frac{\sin 3\theta}{\sin\theta} - \frac{\cos 3\theta}{\cos\theta} = 2$

Solution:

R.S → $\frac{\sin 3\theta}{\sin\theta} - \frac{\cos 3\theta}{\cos\theta} = \frac{\cos\theta \sin 3\theta - \sin\theta \cos 3\theta}{\sin\theta \cos\theta}$

$$= \frac{\sin(3\theta - \theta)}{\sin\theta \cos\theta} = \frac{\sin 2\theta}{\sin\theta \cos\theta} = \frac{2\sin\theta \cos\theta}{\sin\theta \cos\theta} = 2$$

6.4 Inverse of the Trigonometric Functions

Inverse Function	Domain	Range
$y = \sin^{-1} x$	[-1,1]	$[-\pi/2, \pi/2]$
$y = \cos^{-1} x$	[-1,1]	$[0.\pi]$
$y = \tan^{-1} x$	$(-\infty, \infty)$	$(-\pi/2, \pi/2)$

Graph of Inverse Functions:

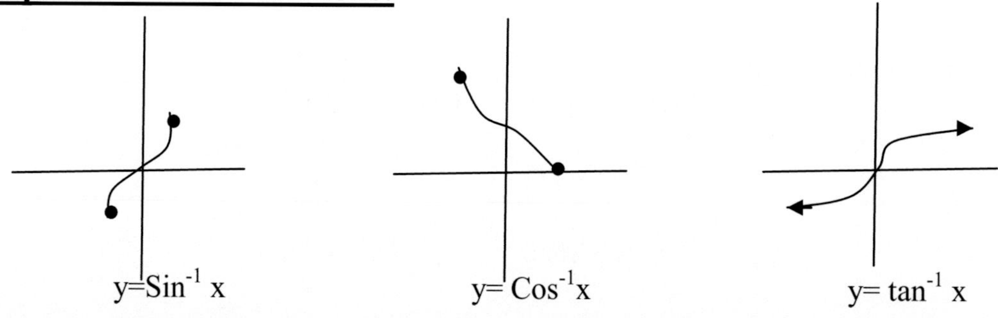

Example-14	Solving Inverse Trigonometric Functions

Find the following inverse function values:
A. Sin^{-1} $(1/\sqrt{2})$
B. $\text{Cos}^{-1}(\sqrt{3}/2)$
C. Sin^{-1} $(-1/2)$
D. Cos^{-1} $(-\sqrt{3}/2)$

Solution:
A. Another way to state the question is: what is the angle θ that has a sine equal to $(1/\sqrt{2})$
And is positive.
Let $\theta = \text{Sin}^{-1}(1/\sqrt{2}) \rightarrow \text{Sin}\,\theta = 1/\sqrt{2} \rightarrow$ Then $\theta=45°$

B. Let $\theta = \text{Cos}^{-1}(\sqrt{3}/2) \rightarrow \text{Cos}\,\theta = \sqrt{3}/2 \rightarrow$ or $\theta = 30°$

C. Let $\theta = \text{Sin}^{-1}(-1/2) \rightarrow \text{Sin}\,\theta = -1/2 \rightarrow \theta= 30°$ in the 4th quadrant or $\theta = 330°$

D. Let $\theta = \text{Cos}^{-1}(-\sqrt{3}/2) \rightarrow \text{Cos}\,\theta = -\sqrt{3}/2 \rightarrow \theta=30°$ in the 3rd quadrant or $\rightarrow \theta=150°$

6.5 Solving Trigonometric Equations

Equation is called a Trigonometric equation if it contains the trigonometric functions;
Example-15: Solve the trigonometric equation: $\sqrt{2}\cos\theta = -1$
Solution: From the given equation $\cos\theta = -1/\sqrt{2}$
From the unit circle θ=45°, we can find the location of this negative cosine which is Either on 2nd, or on the 4th quadrant. If on second quadrant: And the angle is = 135°

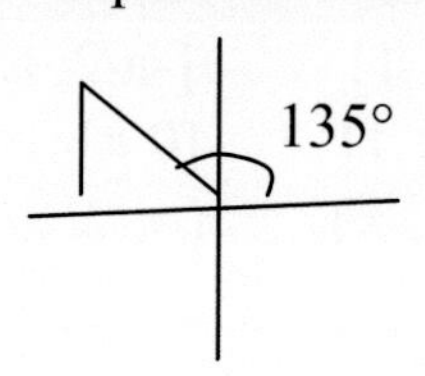

If the angle is on the 3rd quadrant: The angle = 225°

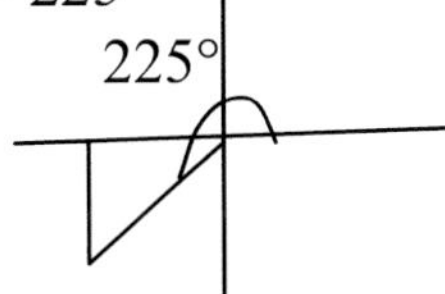

Equation is called a Trigonometric equation if it contains the trigonometric functions;
Example-16: Solve the trigonometric equation: $2 \sin^2\theta = 1 - \sin\theta$ in [0°, 360°]

Solution: Rearranging the equation as:

$2\sin^2\theta + \sin\theta - 1 = 0$, then factoring to:

$(2\sin\theta - 1)(\sin\theta + 1) = 0$

Then: $2\sin\theta - 1 = 0 \rightarrow \sin\theta = ½$ OR $\sin\theta + 1 = 0 \rightarrow \sin\theta = -1$
Solving each separately:
$\sin\theta = ½ \rightarrow \theta = 30°$ and is either on 1st or 2nd quadrant:
On first quadrant $\theta = 30°$, on the second quadrant $\theta = 120°$

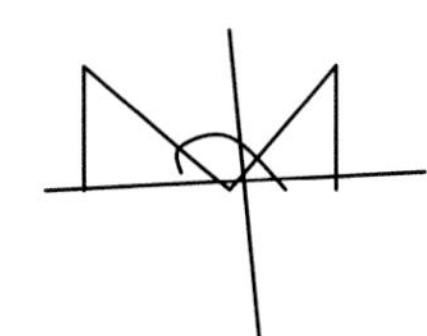

$\sin\theta = -1, \theta = 270°$
The solution is: {30°, 120°, and 270°}

Using TI-83

1. Clear the calculator 2nd → + → 7 → 1 → 2
1. Set the Mode to degrees (or radian).
2. Enter the L.S of the equation as: $y_1 = 2\sin^2\theta$
   Enter the R.S of the equation as: $y_2 = 1 - \sin\theta$
2. Set the window to: x[0,2π], x-scale=π/4, y=[-3,3], y-scale=1
3.

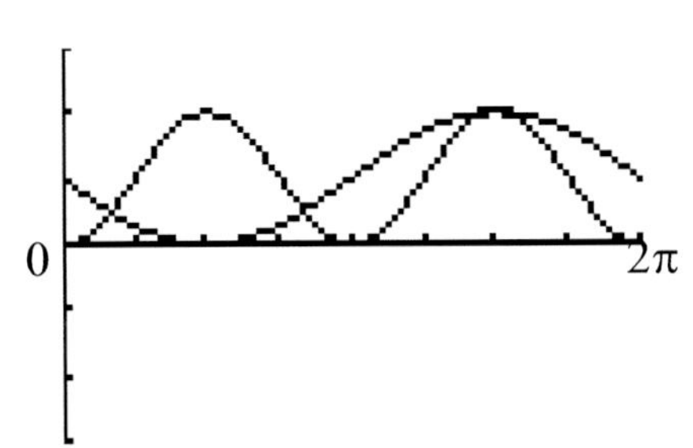

Example-17: Solve the trigonometric equation: $2\cos^2\theta - 5\cos\theta + 2 = 0$, in [0°, 360°)
Solution: First we factor the trigonometric equation: $(\cos\theta - 2)(2\cos\theta - 1) = 0$
Then → $\cos\theta = 2$ or $\cos\theta = 1/2$
→ $\cos\theta = 2$ has no solution, since its range is [-1,1]
→ $\cos\theta = ½ \rightarrow \theta = \cos^{-1}(1/2) = 60°$, and $\theta = 300°$.

Example-18: Solve the trigonometric equation: $\sin^2\theta + 2 \sin \theta \cos \theta + \cos^2\theta = 1$, in $[0°, 360°)$
Solution: we simplify the equation first: $\sin^2\theta + \cos^2\theta + 2 \sin \theta \cos \theta = 1$
But $\sin^2\theta + \cos^2\theta = 1$ → the equation becomes: $2 \sin \theta \cos \theta = 0$ → $\sin \theta \cos \theta = 0$
→ Then $\sin \theta = 0$, or $\cos \theta = 0$ → then the solutions are $\theta = \{ 0°, 90° \}$

Chapte - 6 Exercise

1. Solve the following trigonometric equations in $[0°, 360°\}$

A. $\tan \theta \sin \theta - \tan \theta = 0$
B. $\text{Cos } 2\theta - \sin \theta = 1$
C. $2\sin^2 \theta = 1$
D. $2\sin \theta + \sqrt{2} = 0$
E. $2\cos^2 \theta = \sqrt{3}$
F. $\text{Sec}^2 \theta = 4$
G. Solve for $0° \leq \theta < 360°$: $\tan \theta = 2 \sin \theta$

Chapte - 6 Test

1. Solve exactly for $0 \leq x < 2\pi$: $\cos(2x) + 7\cos(x) = 3$
2. Is the equation an identity? $x^2 + 10x - 100 = x + 10$
3. Solve the following trigonometric equations in $[0°, 360°\}$
   A. $\cos\theta \sin 2\theta + \cos \theta = 0$
   B. $2 \sin^2 \theta + 7 \sin\theta = 4$

7. Application of Trigonometry

Chapter – 7
Application of Trigonometric Functions

Objectives: 7.1 The Law of sine
7.2 The Law of Cosine

Using right triangles we managed to solve them by applying the trigonometric rules , but for non right triangles the trigonometric rules cannot be applied in many cases. Two laws will help us solve these triangles:
1. The Law of Sine's.
2. The Law of Cosine.

7.1 The Law of Sine

The Law of Sine's	For any triangle with different sides: a, b, and c and opposite angles α , β, and γ respectively, then: $\frac{\text{Sin } \alpha}{a} = \frac{\text{Sin } \beta}{b} = \frac{\text{Sin } \gamma}{c}$

Proof: For the given triangle with different sides a, b, c and opposite angles α , β, and γ Respectively:

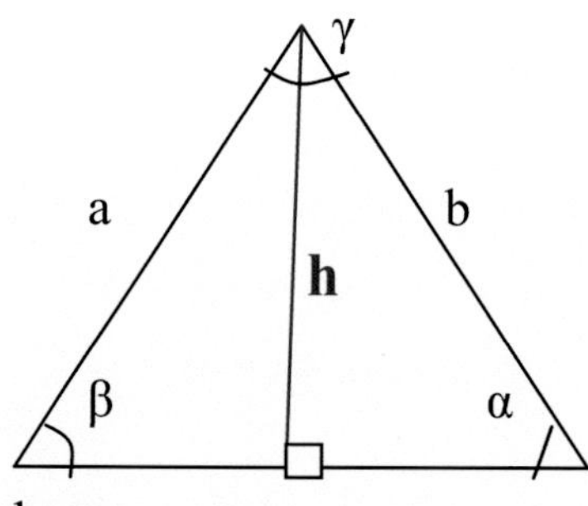

Let h divide the triangle into 2-right triangles as shown.

Now we can use the laws of trigonometry applied to the right triangles as was done before:

$$\text{Sin } \alpha = \frac{\text{Opp}}{\text{Hypo}} = \frac{h}{b} \rightarrow \text{then } h = b \sin \alpha$$

$$\text{Sin } \beta = \frac{\text{Opp}}{\text{Hypo}} = \frac{h}{a} \rightarrow \text{then } h = a \sin \beta$$

From these two relations we get:
$a \sin \beta = b \sin \alpha$ ……(1)

In the similar way if we use h to divide the triangle from different angle into two right triangles:

Now using the same rules:

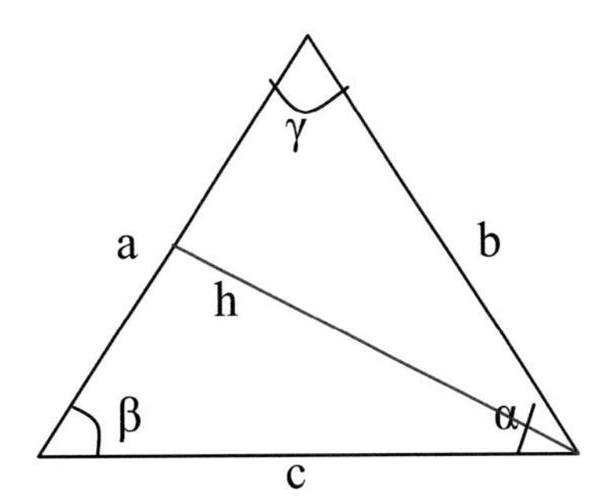

$$\text{Sin } \gamma = \frac{\text{Opp}}{\text{Hypo}} = \frac{h}{b} \rightarrow \text{then } h = b \sin \gamma$$

$$\text{Sin } \beta = \frac{\text{Opp}}{\text{Hypo}} = \frac{h}{c} \rightarrow \text{then } h = c \sin \beta$$

From these two relations we get:
$b \sin \gamma = c \sin \beta$ ……(2)

From (1) → $a \sin \beta = b \sin \alpha$, we get → $\dfrac{\text{Sin } \alpha}{a} = \dfrac{\text{Sin } \beta}{b}$

From (2) → $b \sin \gamma = c \sin \beta$, we get → $\dfrac{\text{Sin } \beta}{b} = \dfrac{\text{Sin } \gamma}{c}$

Then → $\dfrac{\text{Sin } \alpha}{a} = \dfrac{\text{Sin } \beta}{b} = \dfrac{\text{Sin } \gamma}{c}$

Example- 1	**Solving Triangles with given 2 angles and a side AAS**

Solve the triangle with given: a=20 ft, β=45°, γ = 50°
Since 2-angles are give, then we can find the third one:

$\alpha = 180 - (\beta + \gamma) = 180 - (45+50) = 85$

Using the sine law we can get the missing sides:

Then → $\dfrac{\text{Sin } 85}{20} = \dfrac{\text{Sin } 45}{b} = \dfrac{\text{Sin } 50}{c}$ → b = 20 Sin45 / Sin 85 = 14.2 ft
→ c = 20 Sin 50 / Sin 85 = 15.4ft

Type of Triangles

1. SAS If two sides and the angle between them are given

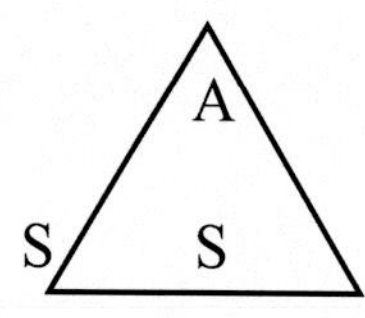

2. SSA If two sides and an angle are given:

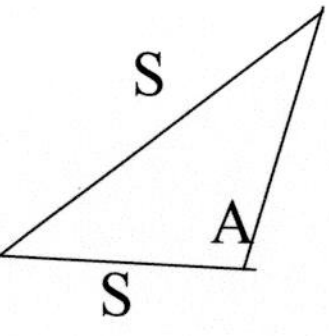

3. SSS If all 3-sides are given

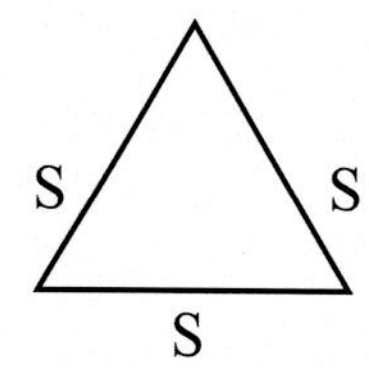

4. AAS If two angles and one side opposite to one of the two angles are given

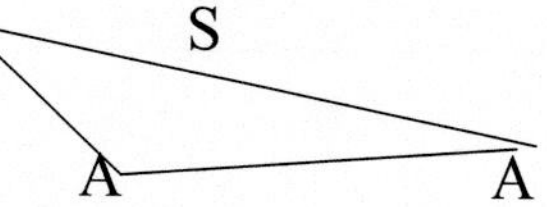

5. ASA If two angles and their side are given

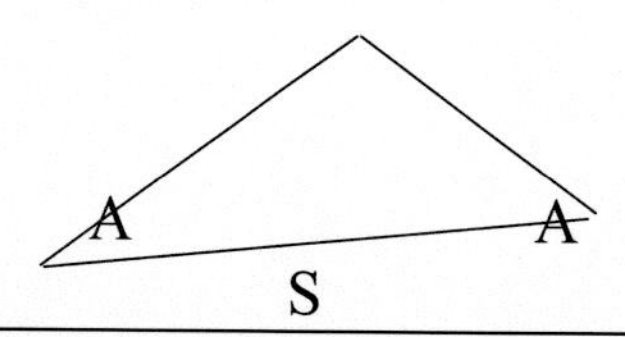

Example- 2	Solving SSA Triangles

Solve the triangle: a=5, b= 7, $\alpha = 34$

Solution: Using the Law of Sine:

$$\frac{\text{Sin } 34}{5} = \frac{\text{Sin } \beta}{7} \rightarrow \text{Sin } \beta = 7/5 \text{ Sin } 34 = 0.78 \text{ then} \rightarrow \beta \approx 31.3 \text{ or } 148.7$$

Since $\alpha + \beta < 180 \rightarrow$ then there are two triangles:

1. With $\beta_1 = 31.3 \rightarrow \gamma_1 = 180 - (\alpha + \beta) = 180 - (34 + 31.3) = 114.7$
2. With $\beta_2 = 148.7 \rightarrow \gamma_2 = 180 - (34 + 148.7) = 2.7$

Now we find the third side using the Law of Sine:

1. $\frac{\text{Sin}\alpha}{a} = \frac{\text{Sin}\,\gamma_2}{c} \rightarrow \frac{\text{Sin } 34}{5} = \frac{\text{Sin } 114.7}{c}$ -→ Solving for $c_1 = 8.12$

2. $\frac{\text{Sin}\,\alpha}{a} = \frac{\text{Sin}\gamma_2}{c} \rightarrow \frac{\text{Sin}34}{5} = \frac{\text{Sin } 2.7}{c} \rightarrow c_2 = 0.42$

The solved triangle is shown below:

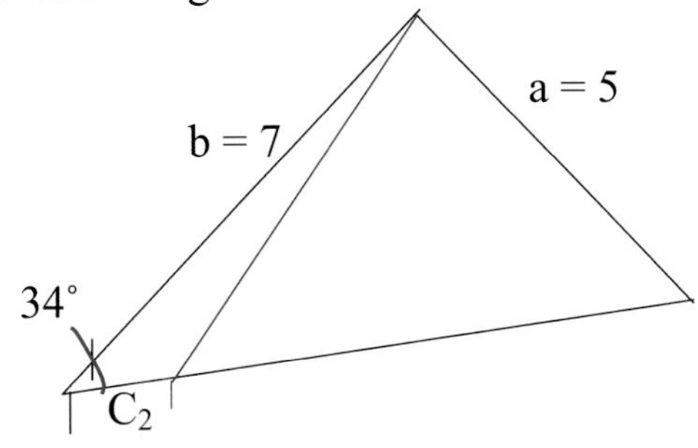

Example-3 | Solving ASA Triangle

Solve the triangle given in the graph:

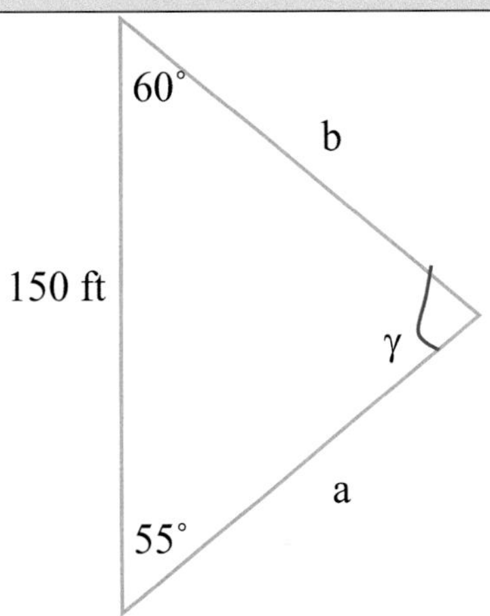

Solution: Using the Sine Law, we need to find the missing angle γ → $\gamma = 180 - (60 + 55) = 65°$

Applying the Law of Sine:

$$\frac{\text{Sin } 60}{a} = \frac{\text{Sin } 55}{b} = \frac{\text{Sin } 65}{150}$$

Then → a = 150 sin 6 / sin 65 = 143.3 ft

And → b = 150 Sin 55 / Sin 65 = 135. 6 ft

Second Method:
This problem can also be solved using the basic trigonometric functions as follows:
Draw the line h normal to the base as shown:
On the upper triangle → $\tan 60 = h/x$ → $h = x \tan 60$…(1)
From the lower triangle → $\tan 55 = h / (150 - x)$ → $h = (150 - x) \tan 55$ …. (2)
From (1) and (2) we get:
$x \tan 60 = (150 - x) \tan 55$ solving for x we get:
$x = 150 \tan 55 / (\tan 60 + \tan 55) = 67.8$ → then $h = x \tan 60 = 117.4$ ft
Solving for a, and b gives → $a = h/\sin 55 = 143.3$ ft, and $b = h / \sin 60 = 135.6$ ft.

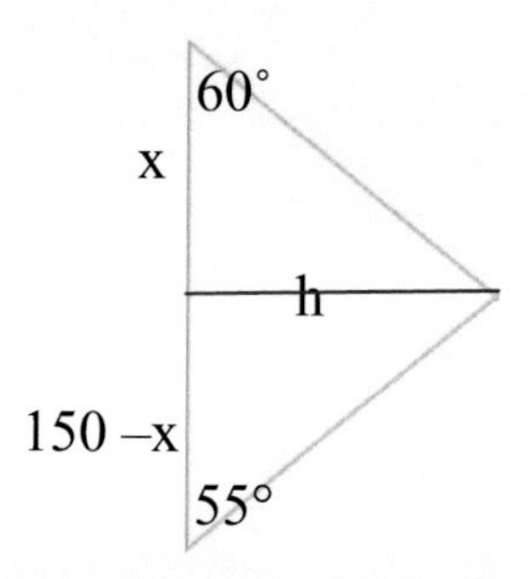

7.2 The Law of Cosines

For a triangle with sides a, b, c and opposite angles α, β, and γ respectively, the law of cosine applies as:
$a = b + c - 2\, b\, c \cos \alpha$ ………..(1)
$b = c + a - 2\, a\, c \cos \beta$ ………..(2)
$c = a + b - 2\, a\, b \cos \gamma$ ………..(3)

Proof: For the given triangle:

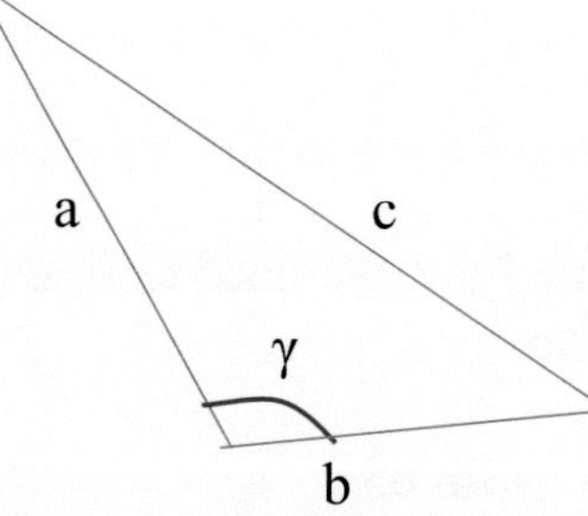

We can extend the base and draw the height of a right triangle as shown in the graph below, then using Pythagorean theorem we can solve for C^2 using the big right triangle :
$C^2 = (d+b)^2 + h^2$
$C^2 = d^2 + b^2 + 2ab + h^2$ → $c^2 = d^2 + h^2 + b^2 + 2ab$

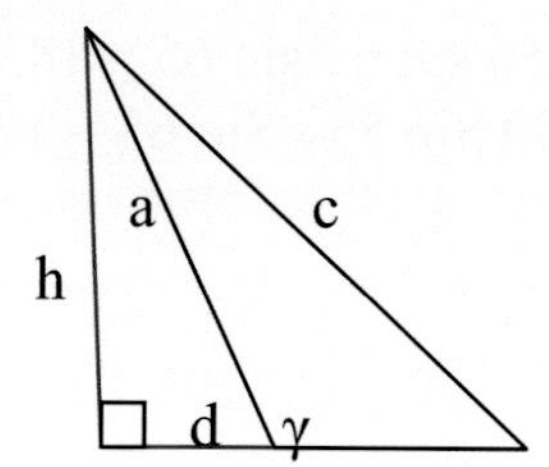

Applying Pythagorean Theorem to the small right triangle:
$a^2 = h^2 + d^2$ substituting in the first equation:
$c^2 = a^2 + b^2 + 2ab$
In the same manner we can prove the rest of the formulas.

Example - 3	Solving SAS Triangle

Solve the triangle given in the figure where $c = 9.3$ feet, $a = 5.4$ feet, and $\beta = 32.3°$

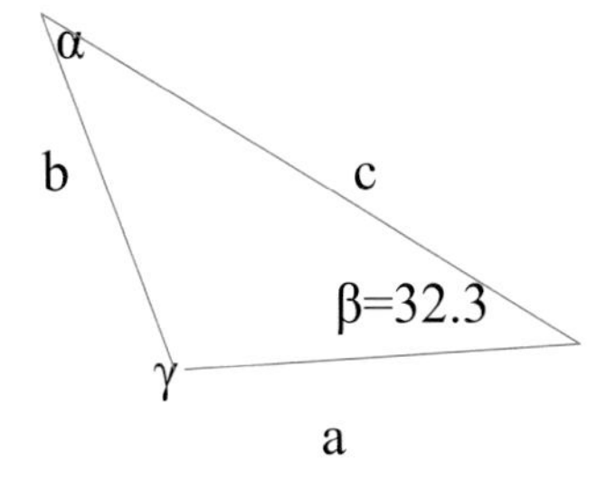

Solution: Using the Law of Cosine:
$b^2 = a^2 + c^2 - 2ac \cos \beta \rightarrow b^2 = (5.4)^2 + (9.3)^2 - 2(5.4)(9.3) \cos (32.3) \rightarrow b = 5.6$ ft
Now using the law of Sine we can find the missing angles:

$$\frac{\text{Sin } 32.3}{5.6} = \frac{\text{Sin } \alpha}{5.4}$$

Then $\rightarrow \alpha = \text{Sin}^{-1}(5.4 \sin 32.3 / 5.6) = 31°$
Then $\gamma = 180 - (32.3 + 31) = 116.7°$

Chapter - 7 Exercise

Solve the following triangles using Sine Law:

1. $\alpha = 73°$, $\beta = 28°$, $c = 42$ feet.
2. $\beta = 83°$, $\gamma = 77°$, $c = 25$ feet.
3. $\beta = 70°$, $\gamma = 10°$, $b = 5$ feet.
4. $\alpha = 110°$, $\gamma = 30°$, $c = 3$.
5. $a = 200$ feet, $\alpha = 32.76°$, $\gamma = 21.97°$.
6. $c = 3$ feet, $\beta = 37.48°$, $\gamma = 32.16°$.

Solve the following Triangles using the cosine law:

7. $a = 5$, $b = 8$, $c = 9$.
8. $b = 1$, $c = 3$, $\alpha = 80°$.
9. $\alpha = 21.2°$, $b=5.32$ ft, $c= 5.05$ ft.
10. $\gamma = 120° 20'$, $\alpha = 5.73$ ft, $b = 10.2$ ft.

Chapter - 7 Test

Solve the following triangles using the Law of Sine:

1. $\alpha = 126.5^\circ$, $a = 17.2$, $c = 13.5$.
2. $\beta = 38^\circ$, $\gamma = 21^\circ$, $b = 24$ ft.
3. $\beta = 43^\circ$, $\gamma = 36^\circ$, $a = 92$ ft.
4. $\alpha = 52^\circ$, $\gamma = 105^\circ$, $c = 47$ ft.

Solve the following Triangles using the Law of Cosine:

5. $a = 2$, $c = 1$, $\beta = 10$
6. $a = 12$, $b = 13$, $c = 5$.
7. $a = 2$, $b = 2$, $\gamma = 50$.

8. Conic Sections

Chapter – 8
Conic Sections

Objectives: 8.1. Introduction
8.2. Distance Formula
8.3. Parabola
8.4. Circles
8.5. Ellipse
8.6. Hyperbola

8.1 Introduction / Conics

In this chapter we will cover the following: The distance between 2 points and deriving the formula for distance between two points; Conic sections such as circles and the method of completing the square to find the standard formula for circles; Parabola, ellipse and graphs. and hyperbola.
The word **Conic** is derived from word cone; conic sections are curves that results from the intersection of plane with a right circular cone which forms parabola opened to the right or left, a circle, an ellipse or hyperbola.

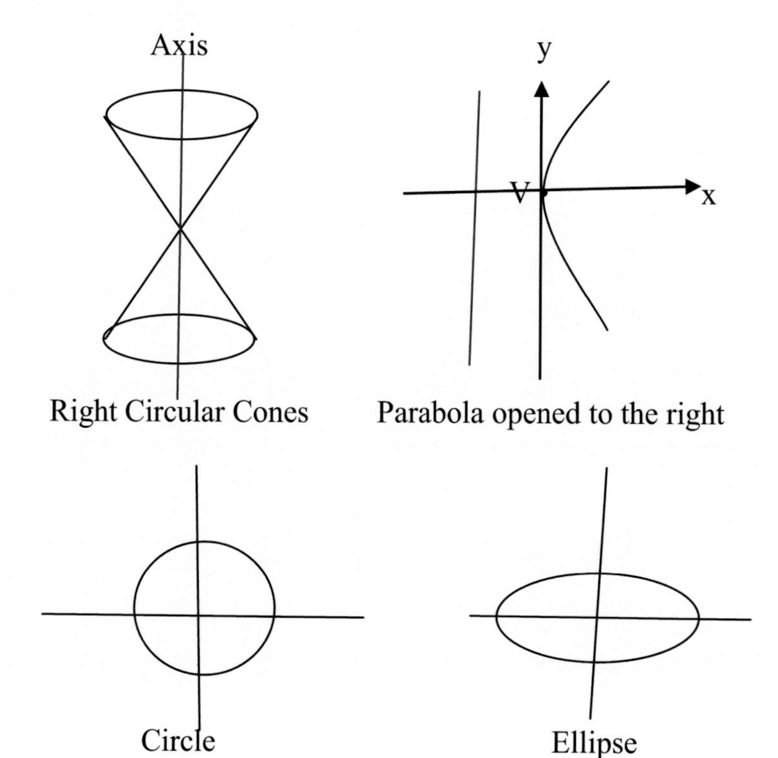

Right Circular Cones Parabola opened to the right

Circle Ellipse

8.2 Distance Formula

The distance d between two points $P_1(x_1, y_1)$, and $P_2(x_2, y_2)$ in the rectangular system can be derived as follows: draw a vertical line at P_2 and a horizontal line at P_1, the two lines will intersect at $P_3(x_2, y_1)$ forming a right triangle. Now we can use Pythagorean Theorem to find the distance d:

Rise = $y_2 - y_1$

Run = $x_2 - x_1$

Then by Pythagorean Theorem:

$d^2 = \text{Run}^2 + \text{Rise}^2$

$d = \sqrt{(x_2 - x_1)^2 + (y_2 - y_1)^2}$

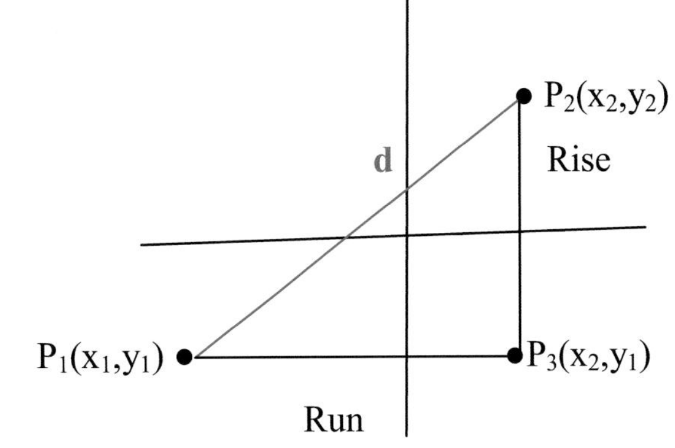

The mid-point of d is $\left(\frac{x_1 + x_2}{2}, \frac{y_1 + y_2}{2}\right)$

Example-1	Find the distance between the two points (-2, -5) and (3,7), and find the midpoint between them

Solution: Using the distance formula:

$$d = \sqrt{(3 - (-2))^2 + (7 - (-5))^2} = \sqrt{25 + 144} = \sqrt{169} = 13$$

$$\text{Midpoint} = \left(\frac{-2+3}{2}, \frac{5+7}{2}\right) = (1/2, 1)$$

Practice-1	Find the distance between the two points (–1/2, –5/4), (1/3, 9/3), and find the midpoint between them

8.3 Parabola

In the previous chapter we have discussed the parabola to be a second degree polynomial, and solved problems, here we will concentrate on the vertex of the parabola denoted by v(h, k). Vertex represents the maximum point if the parabola is opened down and minimum point if the parabola is opened up. To find the vertex we will apply the method of completing the square to the second degree polynomial equation or the quadratic equation:

$f(x) = ax^2 + bx + c$

$= a(x^2 + b/a\ x) + c$

$= a(x^2 + b/a\ x + (b/2a)^2 - (b/2a)^2) + c$

$= a(x^2 + b/a\ x + (b/2a)^2) + c - (b/2a)^2$

$= a(x + b/2a)^2 + (4ac - b^2)/4a$

$f(x) = (x-h)^2 + k$, where $h = -b/2a$, and $k = f(h) = (4ac-b^2)/4a$

Quadratic equation in Vertex form	If $h = -b/2a$, and $k = (4ac-b^2)/4a$, then $f(x) = ax^2 + bx + c = a(x-h)^2 + k$ Where,(h, k)= (– b/2a, f(/b/2a)) is the vertex of the parabola, And the axis of symmetry is the line x= – b/2a

Max, and Min Of parabola **Domain** **Range**	V(h,k) = max point if a < 0 → parabola opened down V(h,k) = min if a > 0 → parabola opened up Domain of all parabolas is all the real numbers d = (–∝,∝) Range is (-∝, k) if parabola is opened down. Range is (k, ∝) if parabola is opened up

Example-2	For the quadratic function: $f(x) = 2x^2 + 8x + 5$ • Find the domain and the range • Rewrite the function in vertex form • Find the vertex • Find the axis of symmetry • Graph f(x)

Solution: The Domain is all the real number

Using completing the square method to find the vertex

$f(x) = 2x^2 + 8x + 5$

$(x) = 2(x^2 + 4x) + 5$

$= 2(x^2 + 4x + (+4/2)^2 + (-4/2)^2) + 5$

$= 2(x^2 + 4x + (4/2)^2) - 2(4/2)^2 + 5$

$f(x) = 2(x^2 + 4x + 4) - 8 + 5$

$f(x) = 2(x + 2)^2 - 8 + 5$

$f(x) = 2(x + 2)^2 - 3 \rightarrow$ this gives $h = -2$, and $k = -3$

The vertex can also be found by using the formula's directly:

$a = 2$, $b = 8$, and $c = 5$ Then $h = -b/2a = -8/2(2) = -2$, and

$k = f(h) = f(-2) = 2(-2)^2 + 8(-2) + 5 = 8 - 16 + 5 = -3$

$V(h,k) = (-2, -3)$, from this we get the equation of the axis of symmetry line is

$h = -2$ or $x = -2$

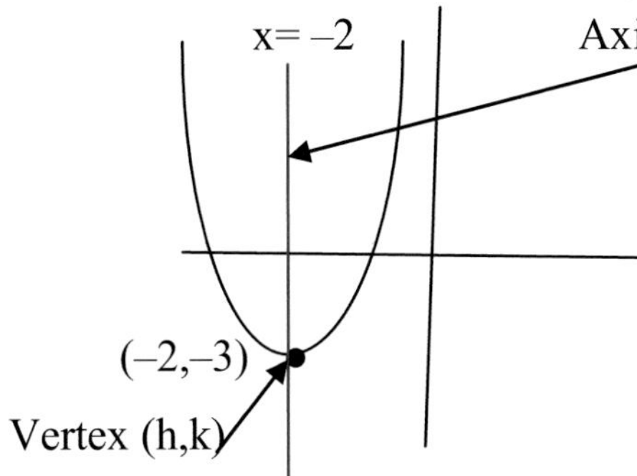

Domain = all the real numbers

Range = $(k,\infty) = (-3,\infty)$

Practice-2	For the quadratic function: $f(x) = -3x^2 + 6x + 1$ • Find the domain and the range • Rewrite the function in vertex form • Find the vertex • Find the axis of symmetry • Graph f(x)

Example-3	For the quadratic function: $f(x) = x^2 - 6x + 9$ • Find the domain and the range • Rewrite the function in vertex form • Find the vertex • Find the axis of symmetry • Graph f(x)

Solution: Since $a > 0$ then the parabola is opened up, with minimum point at the vertex (h,k) and the domain of f(x) is all the real numbers and the range = (k,∞) , we need to find k.

$f(x) = x^2 - 6x + 9$

$= (x^2 - 6x) + 9$

$= (x^2 - 6x + (3)^2 - (3)^2) + 9$

$= (x^2 - 6x + (3)^2) - (3)^2 + 9$

$f(x) = (x^2 - 6x + (-3)^2) - (-3)^2 + 9 \rightarrow = (x - 3)^2 + 0$

$f(x) = (x - 3)^2 \rightarrow$ then h = 3, k=0 this means the parabola touches the x-axis.

Using direct formula will lead to the same solution:

$h = -b/2a = -(-6)/2(1) = 3$, and

$k = f(h) = f(3) = (3)^2 - 6(3) + 9 = 0$

Then axis of symmetry is x=h = 3

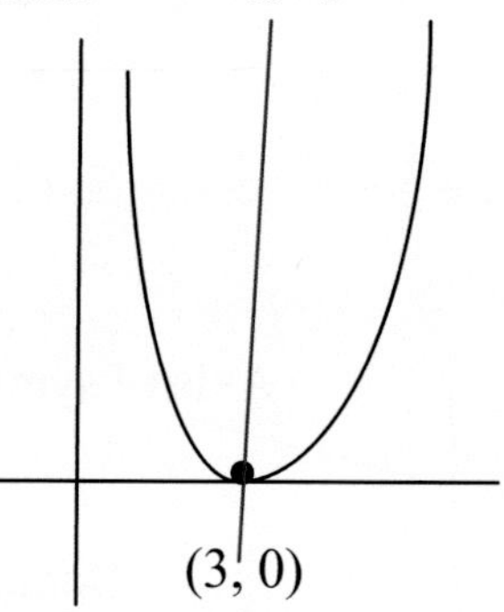

Practice-3	For the quadratic function: $f(x) = 2x^2 + x + 1$ • Find the domain and the range • Rewrite the function in vertex form • Find the vertex • Find the axis of symmetry • Graph f(x)

Example-4	Find the quadratic function from the given information: $V(h,k) = (1,-5)$, and y-intercept = –3

Solution: Using the quadratic function in vertex form:

$f(x) = a(x - h)^2 + k$, substitute the values of h =1, and k = –5

$f(x) = a(x-1)^2 - 5$ use the y-intercept to find a

Since y-intercept point is (0, –3)

Then $\rightarrow f(0) = -3 = a(x-1)^2 - 5$

$-3 = a(x^2 - 2x + 1) - 5$

$-3 = a(0 - 2(0) + 1) - 5$

$-3 = a - 5 \rightarrow a = 2$

Then the function is $f(x) = 2(x-1)^2 - 5$

Practice-4	Find the quadratic function from the given information: V(h,k) = (1,–3), and y-intercept = –2

8.4 Circles

Definition	The circle is defined as a set of points in the xy-plane at a fixed distance r (radius) from the origin if the circle is centered at the origin, or distance r from a point (h,k) if the circle is not centered at the origin

Radius r	The radius r can be found using the distance formula because it is the distance between a point on the circumference and the origin of the circle $r = \sqrt{(x-h)^2 + (y-k)^2}$, (h,k) point at the center If the circle is centered at the origin, then (h,k) =(0,0) and the formula will be: $r = \sqrt{x^2 + y^2}$

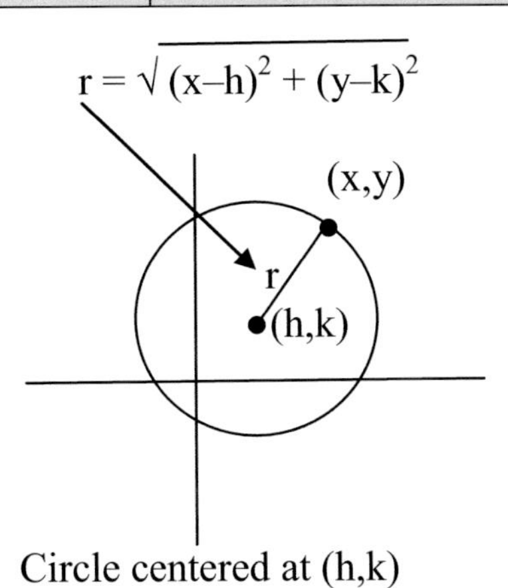

Circle centered at (h,k)

Circle centered at the origin (0,0)

Equation of the circle	Equation of the circle in the standard form: $r^2 = (x-h)^2 + (y-k)^2$ centered at (h,k) And $r^2 = x^2 + y^2$ centered at the origin(0,0)
Equation of the circle in the general form	Equation of the circle in general form is: $x^2 + y^2 + ax + by + c = 0$

Example-5	Find the standard form of the equation of the circle with radius 6 centered at (–3,–5), and graph.

Solution: Given r=6, h= –3, and k= –5, we substitute these values on the standard form:

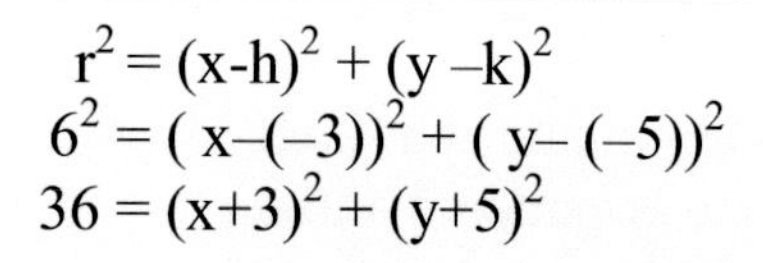

$$r^2 = (x-h)^2 + (y-k)^2$$
$$6^2 = (x-(-3))^2 + (y-(-5))^2$$
$$36 = (x+3)^2 + (y+5)^2$$

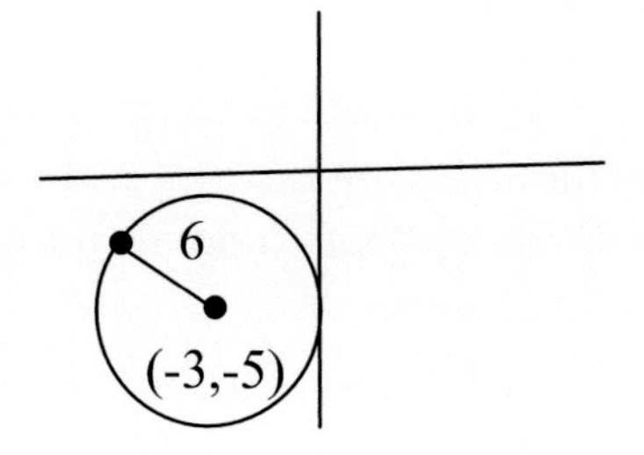

Practice-5	Find the standard form of the equation of the circle with radius 10 centered at (–3, 8), and graph.

Example-6	For the given equation, find the center, the radius and graph the circle: $x^2 + y^2 + 6x + 4y + 9 = 0$

Solution: Group x-terms and y-terms separately, and take the constant to the R.S:

$(x^2 + 6x) + (y^2 + 4y) = -9$ complete the square of each group

$$(x^2 + 6x + (6/2)^2 - (6/2)^2) + (y^2 + 4y + (4/2)^2 - (4/2)^2) = -9$$
$$(x^2 + 6x + 9) + (y^2 + 4y + 4) = -9 + 9 + 4$$
$$(x+3)^2 + (y+2)^2 = 4$$

Then the circle is centered at (h,k) = (–3, –2) with radius r =2

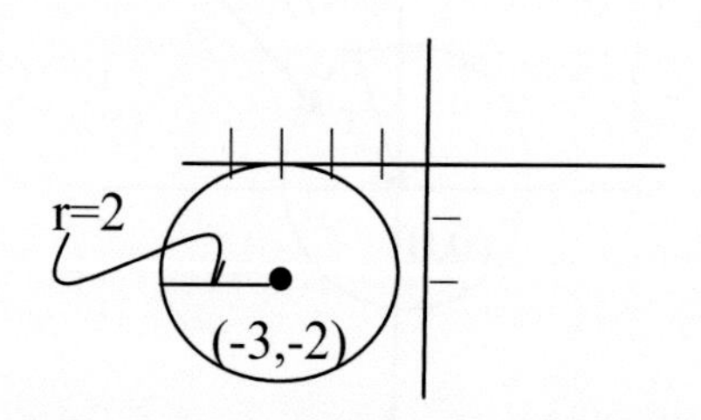

Practice-6	For the given equation, find the center, the radius and graph the circle: $x^2 + y^2 - 2x - 4y - 4 = 0$

8.5 Ellipse

The ellipse is a collection of points in the plane. It has two major points called **foci** which are fixed (F_1, F_2), and two axis (major, and minor)

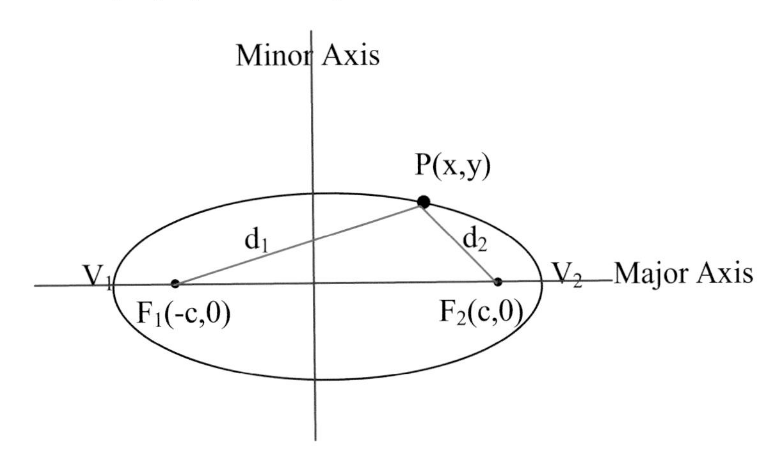

For any point on Ellipse $d_1 + d_2$ = constant = 2a

Equation of Ellipse With center at the origin **Horizontally**	Equations of the ellipse with center at the origin (0,0) and foci at $F_1(-c,0)$ and $F_2(c,0)$ with vertices $V_1(-a,0)$ and $V_2(a,0)$, with the minor axis at (0,b) and (0,-b): $\frac{x^2}{a^2} + \frac{y^2}{b^2} = 1$ where, $a > b > 0$, and $b^2 = a^2 - c^2$

Vertically	$\frac{x^2}{b^2} + \frac{y^2}{a^2} = 1$ where, $a > b > 0$, and $b^2 = a^2 - c^2$
Equation of Ellipse With center at (h,k)	Equation of the Ellipse at (h,k) is:
Horizontally	$\frac{(x-h)^2}{a^2} + \frac{(y-k)^2}{b^2} = 1$
Vertically	$\frac{(x-h)^2}{b^2} + \frac{(y-k)^2}{a^2} = 1$

Finding the Equation of the Ellipse

Applying the formula: $d_1 + d_2 = 2a$, then using distance formula gives,

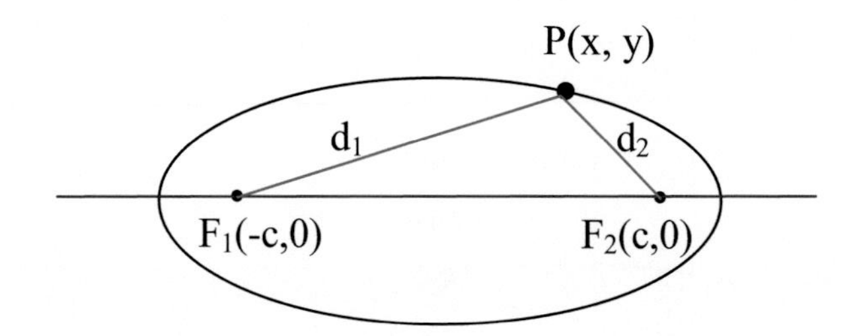

$d_1 + d_2 = 2a$

$d_1 = \sqrt{(x+c)^2 + y^2}$, and $d_2 = \sqrt{(x-c)^2 + y^2}$, then,

$\sqrt{(x+c)^2 + y^2} + \sqrt{(x-c)^2 + y^2} = 2a$

$\sqrt{(x+c)^2 + y^2} = 2a - \sqrt{(x-c)^2 + y^2}$ Squaring both sides,

$(\sqrt{(x+c)^2 + y^2})^2 = (2a - \sqrt{(x-c)^2 + y^2})^2$

$(x+c)^2 + y^2 = 4a^2 + (x-c)^2 + y^2 - 4a\sqrt{(x-c)^2 + y^2}$

$x^2 + 2xc + c^2 + y^2 = 4a^2 + x^2 - 2xc + c^2 + y^2 - 4a\sqrt{(x-c)^2 + y^2}$

$4a\sqrt{(x-c)^2 + y^2} = 4a^2 - 4xc$ Dividing by 4 and squaring both sides,

$(a\sqrt{(x-c)^2 + y^2})^2 = (a^2 - xc)^2$

$a^2 (x^2 -2xc + c^2 + y^2) = a^4 - 2a^2xc + (xc)^2$

$a^2x^2 - 2a^2xc + a^2c^2 + a^2 y^2 = a^4 - 2a^2xc + x^2c^2$

$a^2x^2 - x^2c^2 + a^2 y^2 = a^4 - a^2c^2$

$x^2 (a^2 - c^2) + a^2y^2 = a^2 (a^2 - c^2)$ But $b^2 = a^2 - c^2$

$b^2 x^2 + a^2y^2 = a^2 b^2$ dividing both sides by $a^2 b^2$ Gives,

$\frac{x^2}{a^2} + \frac{y^2}{b^2} = 1$ Equation of the ellipse at the center in a horizontal position

Example-7	Find the equation of ellipse with one focus at (2, 0), and the vertex at (-3, 0).

Solution: from the given information we get:

$c = 2,\ a = 3 \rightarrow$ then $b^2 = (3)^2 - (2)^2 = 5$

The equation of the ellipse is: $\dfrac{x^2}{9} + \dfrac{y^2}{5} = 1$

Practice-7	Find the equation of ellipse with one focus at (4, 0), and the vertex at (– 6, 0), and graph the equation.

Example-8	Find the equation of ellipse with center at (2, –2), vertex at (7,–2) and focus at (4,–2).

Solution: from the given information we get:
Here the ellipse is not centered at the origin but at (h,k)= (2,–2). Since all center, vertex, and focus have y= –2, then the major axis is y= –2 or parallel to x-axis.
The distance from center (2,–2) to focus (4,–2) = 4 –2 = 2 → or c=2.
The distance from the center (2,–2) to vertex (7,–2) = 7–2 = 5 → or a=5
Then $b^2 = a^2 - c^2 = 25 - 4 = 21$, now using the formula for the horizontal ellipse:

$\dfrac{(x-h)^2}{a^2} + \dfrac{(y-k)^2}{b^2} = 1$ we get the equation of the ellipse as:

$\dfrac{(x-2)^2}{25} + \dfrac{(y+2)^2}{21} = 1$ we get the equation of the ellipse as:

Practice-8	Find the equation of ellipse with center at (–3, 1), vertex at (–3, 3) and focus at (–3, 0). Graph the equation.

8.6 Hyperbola

The Hyperbola is a collection of points in the plane. It has two major points called **foci** which are fixed (F_1, F_2), and two axis (Transverse, and Conjugate).

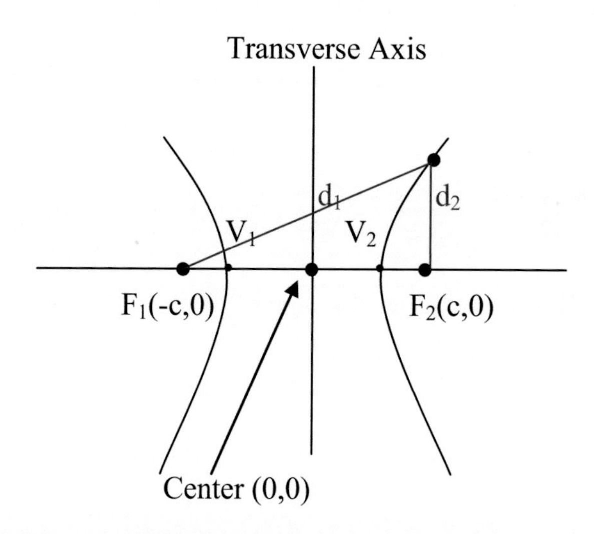

For any point on Hyperbola $|d_1 - d_2| = \text{constant} = 2a$

Equation of Hyperbola With center at the origin	Equations of the Hyperbola with center at the origin (0,0) and foci at $F_1(-c,0)$ and $F_2(c,0)$ with vertices $V_1(-a,0)$ and $V_2(a,0)$, with the minor axis at (0,b) and (0,-b):
Horizontally	$\frac{x^2}{a^2} - \frac{y^2}{b^2} = 1$ where, $a > b > 0$, and $b^2 = c^2 - a^2$
Vertically	$\frac{y^2}{a^2} - \frac{x^2}{b^2} = 1$ where, $a > b > 0$, and $b^2 = c^2 - a^2$

Equation of Hyperbola With center at (h,k)	Equation of the Hyperbola at (h,k) is:
Horizontally	$\frac{(x-h)^2}{a^2} - \frac{(y-k)^2}{b^2} = 1$
Vertically	$\frac{(y-h)^2}{a^2} - \frac{(x-k)^2}{b^2} = 1$

Example-9	Find the equation of hyperbola with center at the origin, one focus at (4, 0), and one vertex at (–1, 0).

Solution: From the given information we have:

One focus is at (3, 0) = (c, 0) → c =3

One vertex is at (–1, 0) = (–a, 0) → a = 1

Then → $b^2 = c^2 - a^2 = 9 - 1 = 8$

Equation of the hyperbola is:

$$\frac{x^2}{1} - \frac{y^2}{8} = 1$$

Practice-9	Find the equation of hyperbola with center at (0, 0), vertex at (1, 0) and focus at (3, 0). Graph the equation.

Example-10	Find the equation of hyperbola with center at (2, –2), vertex at (4,–2) and focus at (5,–2).

Solution: Here the hyperbola is not centered at the origin but at (h,k) = (2,–2)

y= –2 for the center, focus, and vertex this means they lie on axis that is Parallel to x-axis.

The distance from center (2,–2) to the focus (5,–2) = 5–2 = 3 → c=3

The distance from center (2,–2) to the Vertex (4,–2) = 4–2 = 2 → a=2

Then $b^2 = c^2 - a^2 = 9 - 4 = 5$.

→ Equation of the hyperbola is:

$$\frac{(x-2)^2}{4} - \frac{(y+2)^2}{5} = 1$$

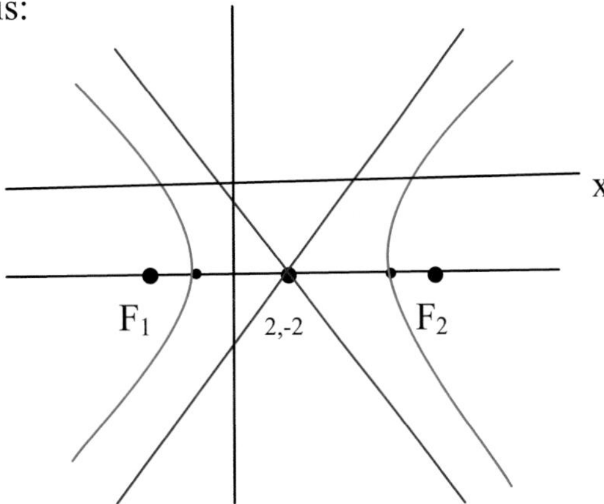

Practice-10	Find the equation of hyperbola with center at (1, 3), vertex at (0, 3) and focus at (–2, 3). Graph the equation.

Chapter - 8 Exercise

Find the distance between each pair of points and find the mid points:

1. (2,5, 1.2) and (1.5, – 5.4)
2. $(-\frac{1}{3}, -\frac{1}{6})$ and $(\frac{3}{4}, \frac{5}{6})$
3. $(2\sqrt{2}, 5\sqrt{6})$ and $(-5\sqrt{2}, 4\sqrt{6})$

For the given center and radius of the circle, write the equation in the standard form:

4. Center at (0, 0) and r=5
5. Center at (–3, 7) and r= 4
6. Center at (– 7, -2) and $r = \sqrt{3}$

Give the center and the radius of the circle described by the given equation:

7. $x^2 + y^2 = 4$
8. $(x-2)^2 + (y-3)^2 = 1$
9 $(x-5)^2 + (y+4)^2 = 16$

10. Complete the square and write the equation in the standard form:
    $x^2 + y^2 + 12x - 6y = 4$

11. Graph the ellipse: $\frac{(x+2)^2}{16} + (y-2)^{2=1}$

12. Graph the ellipse: $4(x-1)^2 + (y-3)^2 = 4$

Chapter - 8 Test

1. Find the distance between the two points and give their midpoint:
   (– 7/2, 10/3) and (– 11/2, –7/3)
2. For the given center and radius of a circle, find its equation in the standard form:
   Center at (-3, 6) and radius = 6
3. Find the center and the radius of the given equation:
   $(x+3)^2 + (y+6)^2 = 9$
4. Complete the squares to find the center and radius of the circle:
   $x^2 + y^2 = 2x + 15$

5. Graph the ellipse: $\frac{(x+2)^2}{16} + (y-2)^{2=1}$

6. Graph the ellipse: $9(x-1)^2 + 4(y-3)^2 = 36$

7. Graph the hyperbola: $9x^2 = 9 - y^2$

Solution to Chapter Exercises

Chapter- 1 Exercise Solutions Page -19

1. The relation is a function
   Domain: {Bob, Ann, Dave}
   Range :{ Ms. Lee, Mr. Bar}
2. The function is not one-to-one function
3. Not one-to-one function.
4. No
5. The inverse of the function={4,-3), (5,-1), 2,0), (4,2), (7,5)}
   Domain = {4, 5, 2, 7}. Range= {-3, -1, 0, 2, 5}.
6. $f^{1}(x) = 1/3(2x-1)$
7. $f^{-1}(x) = (x+8)^{1/3} - 2$
8. The function is $y= (-x+2)^{1/2} - 6$
9. The function is a parabola opened up (a >0), then the domain is D={all the real numbers}
   And the range is from the vertex up, so we need to find the vertex:
   $V = (h,k)$, $h= -b/2a = -(-6) / 2(1) = 3$, and $k=f(h) = (3)^2 - 6(3) +8 = -1$
   Then the range of the function is $R=\{ y/y \geq -1\}$
10. $(-3, -2) \cup (2, 3)$
11. $[-2, 3]$
12. Degree of p(x) is = 6

Chapter- 2 Exercise Solutions Page - 33

1. Quotient = $3x^2 + 11x + 32$, remainder = 99
2. Quotient = $2x^3+2$, remainder = 0
3. Quotient = $x^4 + 3x^3 + 2x^2 +3x +3$, Remainder=3
4. Synthetic Division gives remainder $r = -4$, and $f(-2) = -4$
5. Synthetic division gives remainder r=23, and $f(2) = 23$
6. Zeros are: -1, $3/2$, 5 ; $F(x) = (2x-3)(x-5)(x+1)$
7. Solution set is: $\{(-\infty, -1) \cup (3,\infty)\}$
8. Solution set is $\{[2, 4]\}$
9. Solution set is $\{[6, \infty)\}$
10. a) 120 feet, b) t = 1.38s, or t = 3.62 s (use quadratic formula)
11. Width = 12 yard, Length = 15 yards, Fence = 54 yards.

Chapter- 3 Exercise Solutions Page - 41

1. $(-2 , 4]$
2. $\{ (-\infty, -1) \cup [1,2]\}$
3. $\{(-2, 2) \cup (10, \infty)\}$
4. $[-12,1)\cup[2,\infty)$
5. $(-4, 31]$
6. $ 360
7. x = 120 bicycles.

Chapter- 4 Exercise Solutions Page - 56

1. a) yes b) No 2. No 3. a) No b) yes c) yes 4. Yes 5. Yes

6. yes 8. $f^{-1} = \dfrac{x+1}{x-2}$ 9. $f^{-1} = x^3 - 4$ 10. $f^{-1} = (x-4)^2$

11. $\text{Log}_4 32 = 5/2$ 12. $b^3 = 125$ 13. a. {-2) b. {12}

14. $\text{Log}_2 17 + ½ \text{Log}_2 m - \text{Log}_2 n$ 15. $\text{Log}_6\ x^3 (x-6)^5$ 16. {5}

Chapter- 5 Exercise Solutions Page - 82

1. a =215cos39 = 167.09 ft
   b = 215 sin39 = 135.3 ft

2. $\theta = \tan^{-1}(8/4) \approx 76°$

3. $\alpha = \cos^{-1}(45/60) = 41.4°$

4. Slope of the line m= 2/3 is the tangent of the angle α, then $\alpha = \tan^{-1}(2/3) = 33.7°$

6. The reference angle α = 60.

7. α = π/4

8. α = –√114/6

9. 211.9 ft

10. x=4, y=-12 → r = √160, then the six trigonometric functions are:
    Sin θ = y/r = -12/√ 160
    cos θ = x/r = 4/√ 160
    Tan θ = y/x = -12/4 - -3
    Cot θ = 1/tan θ = -1/3
    Sec θ = 1/cos θ = √ 160 / 4 = √ 10
    Csc θ = 1/sin θ = - √160/ 12 = √10/3

Chapter- 6 Exercise Solutions Page - 92

A. 0, π B. 0, π, 7π/6, 11π/6 C. 30°, 120°
D. 225°, 315° E. 45°, 150°, 210°, 300° F. 60°, 300° G. 0, 60°, 300°

Chapter- 7 Exercise Solutions Page - 99

1. $\gamma = 79^\circ$, a = 41 ft, b = 20 ft.
2. $\alpha = 20^\circ$, b = 25 ft, a = 8.8 ft.
3. $\alpha = 60^\circ$, a = 5.24, c = 0.92
4. $\beta = 40^\circ$, a = 5.64, b = 3.86
5. $\beta = 125.27^\circ$, b = 302 ft, c = 138 ft.
6. $\alpha = 110.36^\circ$, a = 5 ft, b = 3 ft.
7. C = 1.69, $\alpha = 65^\circ$, $\beta = 65^\circ$.
8. A = 2.99, $\beta = 19.2^\circ$, $\gamma = 80.8$.
9. A = 6.03, $\beta = 56.6^\circ$, $\gamma = 52.2^\circ$
10. C = 14, $\alpha = 20^\circ\ 40'$, $\beta = 39^\circ$

Chapter- 8 Exercise Solutions Page - 114

1. Distance ≈ 6.68 2. Distance ≈ 1.47 3. Distance ≈ 10.2

4. $x^2 + y^2 = 25$ 5. $(x+3)^2 + (y-7)^2 = 16$ 6. $(x+7)^2 + (y+2)^2 = 3$

7. C=(0,0); r=2 8. C=(2,3); r=1 9. C= (5,2); r=4 10. $(x-6)^2 + (y-3)^2 = 49$

11. (h,k) =(–2,2), a =4, b=1, $v_1 = (-6, 2)$, $v_2 = (2,2)$

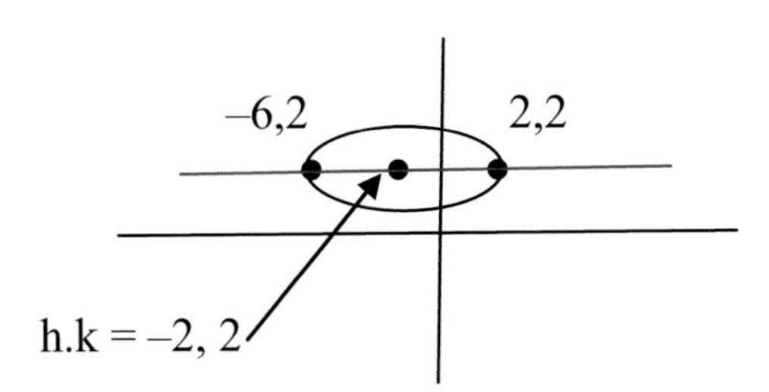

12. (h,k) =(1,3), a =2, b=1, $v_1 = (1, 5)$, $v_2 = (1,1)$

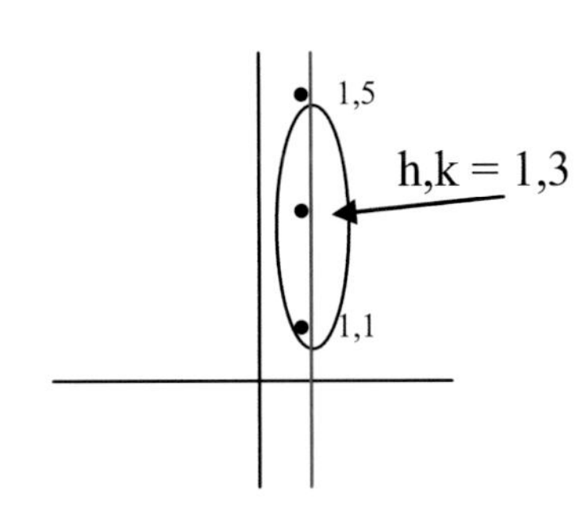

INDEX

E

F

G

H

I

L

M

N

O

P

Z

Information for Reference

Geometry

Rectangle w L	Area $A = WL$	
Square L L	Area $A = L^2$	
Triangle 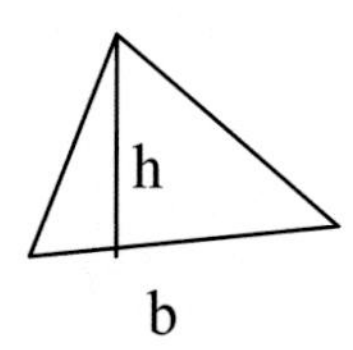 	$A = ½\ hb$	
Right Triangle 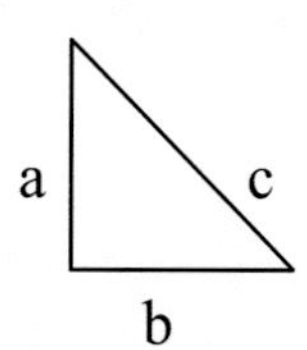 	Pythagorean Theorem $a^2 + b^2 = c^2$	
Measure of angles 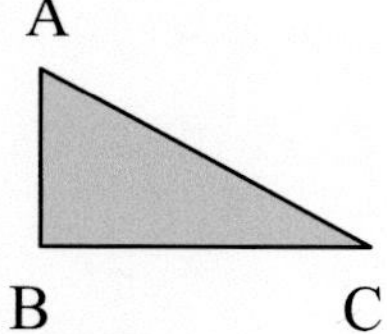 	Sum of angles $A + B + C = 180°$	
Parallelogram 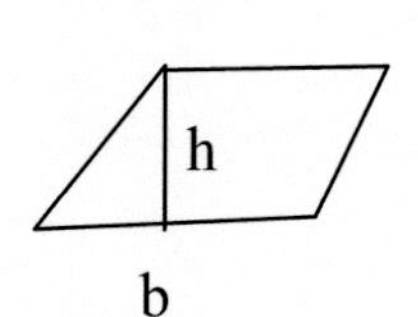 	Area = bh	

Circle:

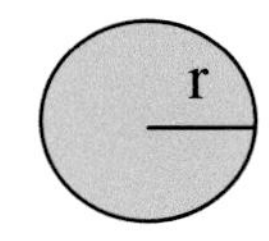

Area $A = \pi r^2$
Circumference $C = 2\pi r$

Cube

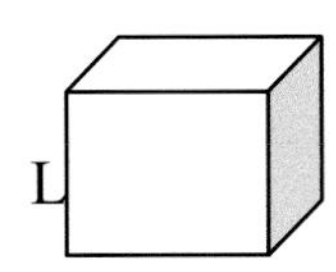

Volume $V = L^3$

Rectangular Solid

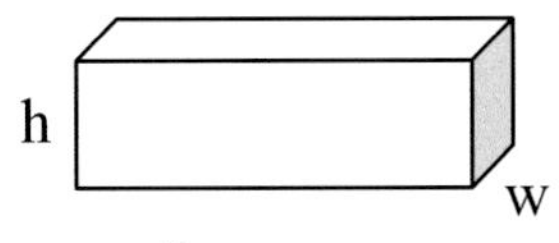

Volume $V = Lhw$

Right Circular Cylinder

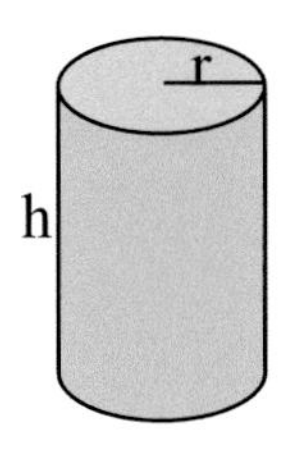

Volume $= \pi r^2 h$
Surface Area $S = 2\pi rh + 2\pi r^2$

Right Circular Cone

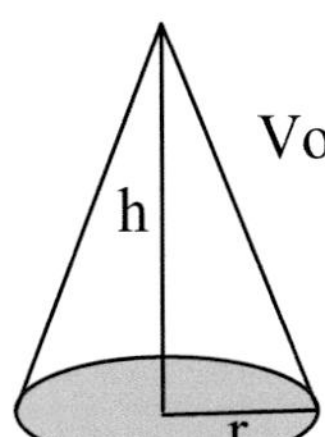

Volume $V = 1/3\ \pi r^2 h$

Trigonometric Identities

1. Reciprocal identities

$$\sin u = \frac{1}{\csc u} \quad \cos u = \frac{1}{\sec u} \quad \tan u = \frac{1}{\cot u}$$

$$\csc u = \frac{1}{\sin u} \quad \sec u = \frac{1}{\cos u} \quad \cot u = \frac{1}{\tan u}$$

2. Pythagorean Identities

$$\sin^2 u + \cos^2 u = 1 \quad 1 + \tan^2 u = \sec^2 u \quad 1 + \cot^2 u = \csc^2 u$$

3. Quotient Identities

$$\tan u = \frac{\sin u}{\cos u} \quad \cot u = \frac{\cos u}{\sin u}$$

4. Co-Function Identities

$$\sin(\frac{\pi}{2} - u) = \cos u \quad \cos(\frac{\pi}{2} - u) = \sin u \quad \tan(\frac{\pi}{2} - u) = \cot u$$

$$\csc(\frac{\pi}{2} - u) = \sec u \quad \sec(\frac{\pi}{2} - u) = \csc u \quad \cot(\frac{\pi}{2} - u) = \tan u$$

5. Even-Odd Identities

$$\sin(-u) = -\sin u \quad \cos(-u) = \cos u \quad \tan(-u) = -\tan u$$

$$\csc(-u) = -\csc u \quad \sec(-u) = \sec u \quad \cot(-u) = -\cot u$$

6. Sum-Difference Formulas

$$\sin(u \pm v) = \sin u \cos v \pm \cos u \sin v$$
$$\cos(u \pm v) = \cos u \cos v \mp \sin u \sin v$$
$$\tan(u \pm v) = \frac{\tan u \pm \tan v}{1 \mp \tan u \tan v}$$

7. Double Angle Formulas

$$\sin(2u) = 2 \sin u \cos u$$
$$\begin{aligned}\cos(2u) &= \cos^2 u - \sin^2 u \\ &= 2\cos^2 u - 1 \\ &= 1 - 2\sin^2 u\end{aligned}$$
$$\tan(2u) = \frac{2 \tan u}{1 - \tan^2 u}$$

8. Power-Reducing/Half Angle Formulas

$$\sin^2 u = \frac{1 - \cos(2u)}{2}$$

$$\cos^2 u = \frac{1 + \cos(2u)}{2}$$

$$\tan^2 u = \frac{1 - \cos(2u)}{1 + \cos(2u)}$$

9. Sum-to-Product Formulas

$$\sin u + \sin v = 2\sin\left(\frac{u+v}{2}\right)\cos\left(\frac{u-v}{2}\right)$$

$$\sin u - \sin v = 2\cos\left(\frac{u+v}{2}\right)\sin\left(\frac{u-v}{2}\right)$$

$$\cos u + \cos v = 2\cos\left(\frac{u+v}{2}\right)\cos\left(\frac{u-v}{2}\right)$$

$$\cos u - \cos v = -2\sin\left(\frac{u+v}{2}\right)\sin\left(\frac{u-v}{2}\right)$$

10. Product-to-Sum Formulas

$$\sin u \sin v = \frac{1}{2}\left[\cos(u-v) - \cos(u+v)\right]$$

$$\cos u \cos v = \frac{1}{2}\left[\cos(u-v) + \cos(u+v)\right]$$

$$\sin u \cos v = \frac{1}{2}\left[\sin(u+v) + \sin(u-v)\right]$$

$$\cos u \sin v = \frac{1}{2}\left[\sin(u+v) - \sin(u-v)\right]$$

Printed in the United States
By Bookmasters